T0350443

This is the first book devoted to quantum state diffusion (QSD) and its applications to open quantum systems and to the foundations of quantum mechanics. It is suitable for readers with basic understanding of quantum theory.

Recent experiments with detailed control over individual quantum systems have changed the face of quantum physics. These systems include atoms at the low temperatures attained by the 1997 Nobel Laureates, they include entangled photons in cavities, and they include the quantum systems used in new and future technologies such as quantum cryptography and quantum computation. The experiments have led to a revival of interest in the foundations of quantum mechanics. QSD is used both as a theoretical and computational tool to study these systems, and as the basis of new approaches to the foundations.

Starting with classical Brownian motion in one dimension, this book leads the reader by easy stages into QSD as the theory of a continuously changing (open) quantum system that interacts with its environment. It then uses the same theory to analyse some modern alternative approaches to the measurement of quantum systems.

The book will interest graduate students and researchers in quantum mechanics and its applications, including quantum optics, quantum statistics, the philosophy of quantum theory and theoretical molecular biology.

QUANTUM STATE DIFFUSION

QUANTUM STATE DIFFUSION

Ian Percival
Queen Mary and Westfield College, London

CAMBRIDGE
UNIVERSITY PRESS

University Printing House, Cambridge CB2 8BS, United Kingdom

Cambridge University Press is part of the University of Cambridge.

It furthers the University's mission by disseminating knowledge in the pursuit of education, learning and research at the highest international levels of excellence.

www.cambridge.org
Information on this title: www.cambridge.org/9780521620079

First published 1998

A catalogue record for this publication is available from the British Library

Library of Congress Cataloguing in Publication data

Percival, Ian, 1931–
Quantum state diffusion / Ian Percival.
p. cm.
Includes bibliographical references and index.
ISBN 0 521 62007 4
1. Quantum theory. 2. Quantum theory–Industrial applications.
I. Title.
QC174.12.P416 1998
530.12–dc21 98-7168 CIP

ISBN 978-0-521-62007-9 Hardback
ISBN 978-0-521-02120-3 Paperback

To Jill

Contents

Acknowledgements

This book could not have been written without generous help from many people and institutions.

I had help from my collaborators and my hosts, Gernot Alber, John Briggs, Tod Brun, Carl Caves, Terry Clark, Predrag Cvitanovic, Lajos Diósi, Barry Garraway, Jonathan Halliwell, Peter Knight, William Power, Peter Richter, Rüdger Schack, Tim Spiller, Walter Strunz and particularly Nicolas Gisin, whose influence can be seen throughout. There were others who willingly provided information or comments on early drafts, including Bobby Acharya, John Charap, Dave Dunstan, Artur Ekert, Gian-Carlo Ghirardi, Lucien Hardy, Frederick Károlyházy, Philip Pearle, Tullio Regge, Antonio Rimini, Steve Thomas, Graham Thompson, Dieter Zeh and many more. Bob Jones, Jagjit Dhaliwal and Nelson Vanegas provided indispensable help in overcoming problems with computers and Steve Adams prepared the figures. Relevant financial support came from the Alexander von Humboldt Foundation, the UK Engineering and Physical Sciences Research Council and the Leverhulme Foundation.

I was lucky to have Simon Capelin, Jo Clegg, Mairi Sutherland and Ian Sherratt of Cambridge University Press, and also Meg Dillon and Jill Percival, to help with the editing and see the book through production, and to have Simon's encouragement and advice.

Thanks to you all.

Ian Percival

1
Introduction

1.1 Description

Recent experiments have changed the face of quantum physics.

They include the detailed control over the states of individual quantum systems such as atoms and photons, entanglement experiments such as tests of Bell's inequalities and also technology based on entanglement, such as quantum cryptography. They include laser cooling to picokelvin temperatures and advances in matter interferometry, leading to measurements of unprecedented precision.

These new experiments need new theories, in particular an analysis of those individual quantum systems which have a significant interaction with their environments, that is individual *open quantum systems*. The experiments expose weaknesses in the usual interpretation of quantum mechanics, and have revived interest in alternative quantum theories. They suggest precision measurements to test alternative quantum theories experimentally and to probe dynamics on Planck scales, such as times of about 10^{-43}s.

This book is about the physics of open quantum systems and the foundations of quantum mechanics. In 1984 Nicolas Gisin made a theoretical connection between them [68, 67], which many authors have developed since then. Quantum state diffusion, or QSD, is a mathematical theory that applies to both.

Questions concerning the theoretical and numerical solution of specific practical problems, like the cooling of atoms to very low temperatures, or the effects of noise on a quantum computer, or the motion of a large molecule like a protein in water, are commonly separated from questions on the foundations of quantum mechanics, like the nature of quantum measurement, or the reality of matter waves in more than three dimensions. In QSD they are closely connected. To many physicists this looks bizarre.

Nevertheless QSD can be applied to both. Applications and foundations have helped one another. Numerical solutions of QSD equations for open systems have provided insight into alternative quantum theories by showing in detail how quantum states can localize during a measurement. And the localization property of QSD, which arose from the study of quantum measurement, is crucial to the efficient numerical solution of some problems of open quantum systems. Experimental tests of alternative quantum theories are difficult because of noise from the environment, which can be analysed using the theory of open quantum systems and the practical application of QSD.

There is a related convergence of two trends in physics previously quite distinct: the quantum measurement problem considered by physicists concerned with quantum foundations, and the quantum measurement process as considered pragmatically by experimenters and their theoretical colleagues looking for intuitive pictures and methods of computation.

The following chapters concentrate on two themes, with little divergence from them: the application of quantum state diffusion to open systems and to the foundations of quantum mechanics. Readers who are interested in related fields will find a guide to the literature at the end of this introduction.

Most physicists working in quantum optics or condensed matter or molecular science naturally do not want to be involved with the foundations. Neither do most physicists or philosophers of science who work on the foundations want to bother too much about numerical and computational methods. The book takes account of this. Also, some of the more mathematical derivations could easily be skipped by readers who are prepared to take the results on trust. The connections between foundations and applications are emphasized here in the introduction, but not in the remaining chapters.

1.2 Summary

Chapters 2 and 3 provide an introduction to the elementary theories of stochastic processes and open quantum systems. The next two chapters on general QSD theory and localization are for all readers. Open systems and quantum measurement are used as examples. Chapter 6 is on QSD as a numerical method for solving open system problems, and provides an introduction to a standard QSD computer program that is available on the World Wide Web. Chapters 7 and 8 are on the foundations of quantum theory. Chapters 9 and 10 contain more specialized theory topics, including classical and semiclassical QSD and the theory of open systems with linear dynamics. The theory in the book is illustrated by many examples.

The dependence of chapters on one another suggests reading sequences given by

$$\begin{array}{l} 1 \rightarrow 2 \rightarrow 3 \rightarrow 4 \rightarrow 5 \rightarrow 6 \\ \qquad\qquad\qquad\quad 5 \rightarrow 7 \rightarrow 8 \\ \qquad\qquad\quad 4 \rightarrow 9 \rightarrow 10, \end{array}$$

but much of the introductory and less mathematical parts can be read independently, particularly in chapters 7 and 8 on quantum foundations.

Chapter 2 introduces open systems, the Itô calculus and the Itô form of Langevin stochastic differential equations from scratch, using Brownian motion as an example. Brownian motion is also used to illustrate the importance of time scales in Markov processes, and to introduce variance and covariance for ensembles. It assumes a nodding acquaintance with elementary probability theory.

Chapter 3 brings in the relevant quantum theory and notation, with the emphasis on open systems, for which interaction with the environment is significant and can be represented using ensembles with probabilities. It requires a background of elementary quantum mechanics in Dirac notation, including the picture of a quantum state as a vector in Hilbert space. Some properties of the density operator are illustrated for two-state systems by motion on and inside the Bloch sphere. The usual definition of a quantum measurement is generalized. Quantum variance and covariance for pure states of individual quantum systems appear.

Chapter 4 uses the methods introduced in the previous two chapters to derive the Langevin-Itô quantum state diffusion (QSD) equations for the quantum states of an individual open system from the master equation for the density operator of an ensemble. Alternative forms of the equations and the connection with quantum jumps are discussed. A gallery of graphs illustrates some solutions for simple examples.

Chapter 5 is on localization. This is the property of QSD that makes the link between classical and quantum mechanics, which provides a dynamics for the process of quantum measurement. It also helps practical computations. Localization is described in terms of variances and entropies which decrease in the mean, and much of the chapter is devoted to showing this. Some readers may want to skip the derivations and just look at the results.

Chapter 6 contains a description of numerical methods for the solution of QSD equations, especially the moving basis, which uses the localization property of QSD. It contains a user's guide to a program library, available on the World Wide Web. The methods are illustrated by some examples, including the Duffing oscillator, second harmonic generation, particles in traps, and quantum computers. The chapter assumes some knowledge of numerical methods and computer modelling, but not the details of the C++ language used for the standard computer library. The results of the computations may be of wider interest.

Chapters 7 and 8 are on the foundations of quantum theory. They present some of the merits and problems of Bohr's Copenhagen interpretation and of those alternative quantum theories that are related to QSD. Some of the material in these chapters assumes an understanding of the physical implications of special and general relativity, but not the detailed theory. The part on experimental tests is written with experimenters in mind.

Chapter 7 gives three reasons for the current interest in alternative theories, one theoretical and two experimental. The difficulties of quantum theory are compared with those of early atomic physics and of the origin of biological species. Bell's conditions for a good quantum theory are stated, and it is shown that neither the Copenhagen interpretation, nor the environmental QSD of the earlier chapters satisfies them. There is a brief account of the development of some good quantum theories that are related to QSD, and a demonstration that some modifications of Schrödinger's equation are difficult to detect, so that there is a great variety of possible good theories, despite an important constraint.

Chapter 8 shows how to restrict the possibilities by appeal to general principles and other fields of physics. Primary state diffusion is derived independently from two very different sets of principles. Firstly it is assumed that state diffusion is primary and the ordinary deterministic Schrödinger evolution is derived from it. This results in energy localization, whose merits and faults are discussed. The second derivation depends on assumed spacetime fluctuations on a Planck scale, due to quantum gravity. Special relativity is applied to the fluctuations, and is needed to produce position localization, but the dynamics of the system remains nonrelativistic. Atom interferometry experiments on spacetime fluctuations and the physics of the classical-quantum boundary are discussed, together with a brief account of the theoretical and experimental problems and prospects.

Chapters 9 and 10 provide a phase space picture of the evolution of individual open quantum systems. Most of it needs a deeper understanding of classical dynamics in phase space than the rest of the book. Those who like to think classically will find a helpful alternative picture of QSD in these chapters. They explain the classical dynamical theory of quantum localization, the semiclassical limit of QSD, and the theory of localized systems with linear dynamics.

1.3 Literature on related fields

This book is about the application of quantum state diffusion to the theory of open quantum systems and to the foundations of quantum theory. It deals either briefly or not at all with many important related fields, for which this section provides a brief guide.

The treatment of quantum measurement in ordinary quantum theory textbooks does not deal adequately with continuous, repeated or incomplete measurements, such as continuous monitoring of quantum systems by photon counting in quantum optics, or nondestructive quantum measurements designed to detect gravitational waves. These continuous measurements can be handled adequately within the framework of ordinary quantum theory, as shown by Davies and Srinivas [26, 142, 141] and Mensky [94]. This provides a basis for the theory of quantum jumps [22, 121] and for the many practical applications of quantum trajectory methods using jumps, which were developed independently, with references given in section 4.7. Continuous measurement theory for gravitational waves is described in [19].

The mathematical theory of the stochastic evolution of quantum states has been developed by Hudson and Parthasarathy [85, 102], Belavkin [9], Barchielli [4] and their collaborators.

John Bell's book [10] provides a stimulating introduction to the foundations of quantum theory. Wheeler and Zurek's collection of classic papers [154] is invaluable. The pilot wave theory of de Broglie and Bohm is treated in two recent books [18, 84]. In his book Bell points to similarities between this theory and the theory of Everett [49, 153]. Bohm discussed the role of causality and probability in quantum mechanics in [16], whereas [93] provides a non-mathematical introduction to the problems of quantum-nonlocality and relativity.

The problems of quantum theory and gravity are mostly well above the level of this book, but the expositions in [108] are clearer than most.

2

Brownian motion and Itô calculus

A Brownian particle is an open classical system, and quantum state diffusion (QSD) is a theory of open quantum systems, so there is a parallel between the physics and mathematics of classical Brownian motion and the physics and mathematics of quantum state diffusion. But Brownian motion is much simpler, so it is used to introduce the properties of open systems, the Itô calculus and the use of diffusion to probe small scales. All of these are important for QSD.

2.1 Brownian motion and QSD

The interaction of an open system like a Brownian particle with its environment significantly affects its dynamics. The motion is stochastic, so the future state of the system is not uniquely determined by its present state. The probability of a given future state *is* uniquely determined. General open systems are defined in section 3.4.

Brownian motion is stochastic motion of an individual open classical system, in contrast to the determinism of Hamiltonian dynamics. Quantum state diffusion is stochastic motion of an individual open quantum system, in contrast to the determinism of Schrödinger dynamics. In QSD the quantum states diffuse in Hilbert space like Brownian particles diffusing in ordinary position space. For example, the system may be an atom in a radiation field, a radiation field in the presence of atoms, a molecule in a liquid, or a quantum computer in the presence of noise. It may be a quantum system which is being measured. In QSD this diffusion is the source of quantum fluctuations and the indeterminism of quantum physics.

Brown did not discover Brownian motion. Anyone looking at water through a microscope is likely to see little things wiggling around, and according to Brown they were seen by Leeuwenhoek, who lived from 1632 to 1723. In 1828 Brown established Brownian motion as an important phenomenon, demonstrated that it was present in inorganic matter and refuted by experiment various spurious mechanical explanations.

In the late nineteenth century there were good scientists like Ostwald and Mach who did not believe in the reality of atoms. For them, atomic theory was just a convenient picture or model which could be used in physics and chemistry to explain the properties of observable macro systems, and those who believed in the reality of atoms were misguided.

Einstein's 1905–1908 theory and the later experiments of Perrin established the reality of atoms without question. Brownian motion is a diffusion process, and because of this, macro scale experiments on Brownian motion could be used to probe the atomic scale, even though the details of atomic dynamics were not known at the time. These experiments are much easier than the more direct methods like X-ray diffraction, scattering experiments and ultramicroscopes. Brown's systematic studies were completed long before such methods were possible. QSD is also a diffusion process. It is suggested in chapter 8 that quantum state diffusion might be used to determine quantities on the Planck scales of length and time, which are smaller than atomic scales by more than twenty orders of magnitude. The role of QSD in quantum theory today parallels the role of Brownian motion in the atomic theory of a century ago.

The books of Einstein and Nelson give good accounts of the early work on Brownian motion [46, 99].

There is a parallel between the theory of Brownian motion and the theory of quantum state diffusion. A Brownian particle moves along a crooked path $\mathbf{r}(t)$ in position space. This crooked path is a solution of a stochastic differential equation, the Langevin equation. Langevin introduced differential equations with stochastic forces, which will be treated more consistently in section 2.3 using Itô's stochastic calculus. The state of an ensemble of Brownian particles is represented by a density $\rho(\mathbf{r}, t)$ that is the solution of a linear deterministic partial differential equation, called the Fokker-Planck equation.

According to QSD, a single open quantum system moves along a crooked path or trajectory $|\psi(t)\rangle$ in the state space, that is a solution of a stochastic differential equation, the QSD equation. The evolution of an ensemble of quantum systems is represented by a density operator $\boldsymbol{\rho}(t)$ that is a solution of a linear deterministic equation, called the master equation, which may be a matrix differential equation or a partial differential equation.

There is also a parallel in the computational methods that are used today. For complicated Brownian processes, such as the motion of charged Brownian particles in an electric field that varies in space and in time, it is often easier to solve the Langevin equation numerically for a sample of particles than to solve the Fokker-Planck equation for the density. The first method is a Monte Carlo method. The Monte Carlo method is less accurate, but it can be used where the Fokker-Planck equation is insoluble in practice. Similarly, for open quantum systems that require many basis states, it is often possible in practice to solve the QSD equation numerically for a sample, when it is impossible to solve the master equation for the density operator, as shown in chapter 6.

These parallels are well established. There are others which are less well established.

For example, the theory of Brownian motion clarified the foundations of kinetic theory, whereas QSD clarifies the foundations of quantum mechanics. Theory and experiments on Brownian motion provided access to the physics of atomic scales, whereas the primary state diffusion theory and matter interferometer experiments of chapter 8 might provide access to the physics of the Planck scales.

So Brownian motion and its theory provide a good introduction and background to the physics and mathematics of QSD, despite the important differences between them. For simplicity, consider only the vertical motion of a Brownian particle.

First suppose that the density of the particle is the same as the fluid around it. Then there is no drift due to gravity. The displacement of the particle fluctuates due to collisions of the Brownian particle with molecules. This process takes place on the atomic scale, producing random upward and downward displacements, resulting in a diffusion away from its initial height.

The mean vertical displacement $x(t)$ of the particle from its initial height is zero,

$$\mathrm{M}\,x(t) = 0, \tag{2.1}$$

where M will always be used to represent the mean over stochastic fluctuations. 'Expectation' will not be used for such fluctuations, as it is needed for quantum expectations. As in statistical mechanics, a mean is a property of a large ensemble of individual systems. For our example, the mean vertical displacement is a property of an ensemble of Brownian particles that start from the same height.

The mean square displacement from the initial position is proportional to the time,

$$\mathrm{M}\, x^2(t) = a^2 t, \tag{2.2}$$

where by convention a is a positive constant. The value of a depends on the mass of the particle, the viscosity of its fluid environment and the temperature, but this does not concern us here. A typical displacement due to the diffusion is proportional to the square root of the time, and this can be expressed in terms of the root mean square (RMS) deviation,

$$\Delta x(t) = \sqrt{\mathrm{M}\, x^2(t)} = a\sqrt{t}. \tag{2.3}$$

When the density of the Brownian particle is greater than the density of the fluid through which it moves, there is an additional drift due to gravity. Its vertical displacement is then made up of two parts, drift and stochastic fluctuations,

$$\text{vertical displacement} = \text{drift} + \text{stochastic fluctuations}. \tag{2.4}$$

Here and throughout the book a *fluctuation* will always be a stochastic variable with zero mean. It is only then that there is a unique decomposition of the motion into separate drift and fluctuation components. Here a fluctuation is a displacement and not a velocity.

The drift, due to the force of the Earth's gravitational field, is uniformly downwards. In a short time interval δt, the mean square displacement $\mathrm{M}\,(\delta x)^2$ due to the diffusion is proportional to δt, so a typical displacement is proportional to the square root of the time interval. The drift is proportional to the time interval δt itself. The absence of the expected acceleration term is explained in section 2.8.

Figure 2.1 illustrates simulated Brownian motion with drift. Figure 2.1(a) gives the height. Over very short periods of time it is difficult to see the drift because it is masked by the fluctuations, whereas over longer periods the drift becomes relatively more important. Over very long periods it may be difficult or impossible to distinguish the fluctuations at all. Figure 2.1(b) shows the root mean square displacement $\sqrt{\mathrm{M}\,(x^2)}$ from the initial height at $t = 0$ for a sample of 100 Brownian particles. This is not a deviation from the mean. For

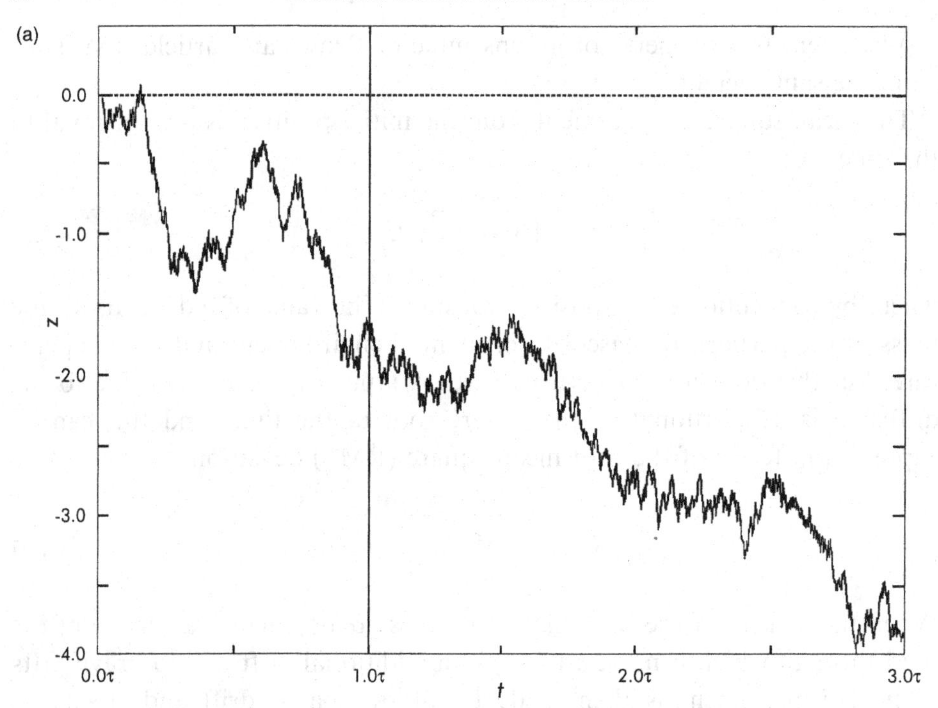

Figure 2.1a The height of a Brownian particle as a function of time.

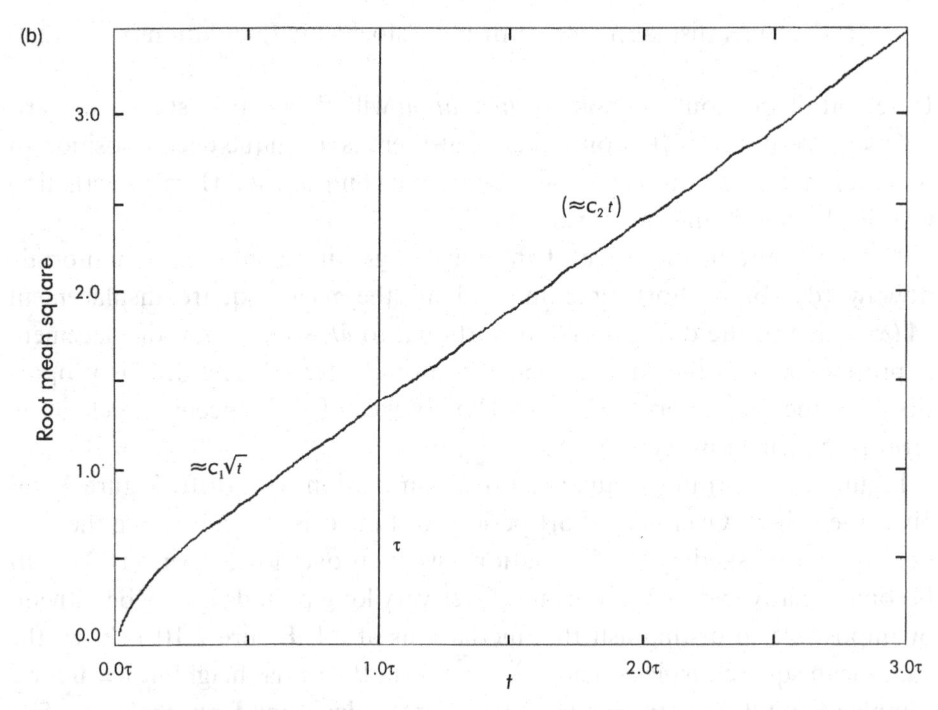

Figure 2.1b The RMS deviation from the initial height for a sample of 100 particles. τ is defined in the text.

small times it looks roughly parabolic, typical of the square root dependence of the fluctuations without the drift. For longer times it looks linear, typical of drift without fluctuations.

The figure also shows the time interval τ that marks the approximate boundary between the smaller time intervals for which the diffusion dominates and the larger time intervals for which the drift dominates. The measured value of τ provides valuable information about atomic scales.

For times T much longer than τ, the fluctuations are typically about $(\tau/T)^{1/2}$ of the displacement due to the drift. This is the *fluctuation factor*, which is defined for all processes in which there are both fluctuations and drift. It is important for the experimental tests of spacetime fluctuations proposed in chapter 8.

2.2 Probabilities

A Brownian particle is a special case of a system interacting with its environment. The state of the environment cannot be represented exactly, but, as usual in statistical mechanics, the probability of a given state can be represented. Consequently, interaction between system and environment produces indeterminate stochastic evolution of the system, represented by an ensemble with a probability distribution in space for the Brownian trajectories.

The notation used here for probabilities is easily extended to quantum systems. Let $\mathbf{Pr}(x)$ be a probability distribution for a discrete or continuous variable x. Then the total probability for any value of x must be 1, so

$$\sum_x \mathbf{Pr}(x) = 1 \qquad \text{or} \qquad \int \mathrm{d}x \cdot \mathbf{Pr}(x) = 1, \tag{2.5}$$

where the sum or integral is taken over all possible values of x. Such sums and integrals are so common that it is convenient to use a general notation for both which is consistent with the corresponding quantum notation. The *trace* Tr is the sum or integral over all possible values of a variable x and so

$$\mathrm{Tr}\,\mathbf{Pr}(x) = 1. \tag{2.6}$$

The variable x can represent a dynamical variable of some classical system, like the height of a Brownian particle. If $f(x)$ is some property of the system, like the vertical distance of the particle from some point, then the mean of f for the probability distribution $\mathbf{Pr}(x)$ is

$$\mathrm{M}f = \mathrm{Tr}\,\mathbf{Pr}(x)f(x). \tag{2.7}$$

The *ensemble variance* or *mean square deviation* of f is

$$\Sigma^2(f) = \mathrm{M}\,(f - \mathrm{M}f)^2 = \mathrm{M}\,(f^2) - (\mathrm{M}f)^2. \tag{2.8}$$

This is always positive or zero and its square root Δf is the standard deviation. All the variances of this chapter are classical ensemble variances. In chapter 3 we will meet the corresponding quantum variances.

The probability distribution for $y = f(x)$ is

$$\mathbf{Pr}(y) = \mathrm{Tr}_x\,\mathbf{Pr}(x)\,\delta(f(x) - y), \tag{2.9}$$

where the δ function is the Dirac or Kronecker δ, depending on whether x is continuous or discrete.

2.3 Itô calculus

Itô and Stratonovich have done for stochastic dynamics what Newton and Leibniz did for ordinary dynamics, by treating the small displacements as differentials. This results in a very elegant theory and differential calculus for continuous stochastic processes that satisfy Langevin equations, sometimes known as Wiener processes. This and other stochastic methods are described for physicists in the books of Gardiner [55, 56]. Remarkably, Itô calculus is a differential calculus of nondifferentiable functions.

When a Brownian particle with vertical displacement x diffuses with no drift, the mean of the vertical distance traversed is zero and the mean square or variance is proportional to the time. So for small displacements $\mathrm{d}x$,

$$\mathrm{M}\,\mathrm{d}x(t) = 0, \qquad \mathrm{M}\,[\mathrm{d}x(t)]^2 = a^2\mathrm{d}t, \tag{2.10}$$

where M represents a mean over all possible displacements $\mathrm{d}x(t)$ and a is a positive constant. The small displacements are treated as differentials by Itô, and their higher powers are neglected, as in the ordinary Leibniz differential calculus, but the rules are different. For more complicated stochastic processes, it is convenient to introduce a standard normalized stochastic differential $\mathrm{d}w$, with zero mean and variance *equal* to the time, so that

$$\mathrm{M}\,\mathrm{d}w = 0, \qquad \mathrm{M}\,(\mathrm{d}w)^2 = \mathrm{d}t. \tag{2.11}$$

For these differential fluctuations, only these means are significant. It does not matter whether the ensemble of stochastic differentials is made up of steps of fixed length with equal probability of going to the left or to the right, or of steps with a Gaussian distribution, or a rectangular distribution. They all have the same effect, provided that they satisfy (2.11).

It is implicitly assumed that fluctuations at different times are independent of one another. This is the Markov assumption, discussed in more detail in section 2.8. Brownian motion with no drift is then given by the solution of the simplest Langevin-Itô differential equation

$$\mathrm{d}x(t) = a\,\mathrm{d}w(t). \tag{2.12}$$

In this equation, and in general for Itô stochastic differential equations, it is assumed that, at time t, the trajectory $x(s)$ with $0 \leq s \leq t$ is already determined, so that only the new fluctuation $\mathrm{d}w(t)$ need be considered. All the particles of the ensemble are at the initial point $x(t)$ at time t, but they are usually at different points at time $t + \mathrm{d}t$. So for a single step of the equation the mean M refers to the ensemble of the current fluctuations $\mathrm{d}w(t)$ only. Functions of t and $x(t)$ have a fixed value, and so may be taken outside the mean.

The properties of more general ensembles, formed from all the fluctuations over finite time intervals, can be derived from the properties of these elementary ensembles, in which the initial ensemble consists of systems in the same state. That is why most of the Itô equations in this book have no mean M on the right hand side. There is no need for it when all the initial states are identical, and general equations can be obtained from the elementary ensembles.

In ordinary differential calculus the whole theory can be based on first-order differentials, higher orders being neglected by comparison with the first order. The Itô differential calculus is also based on lower-order differentials, but they are not just of first order. Stochastic differentials are of order $(\mathrm{d}t)^{1/2}$, so the squares of stochastic differentials like $\mathrm{d}w$ cannot be neglected.

Squares of ordinary differentials like $\mathrm{d}t$ can be neglected as usual, and so can products of the form $\mathrm{d}t\,\mathrm{d}w$, which are of order $(\mathrm{d}t)^{3/2}$. The contributions of drift and diffusion terms can be considered independently, because cross terms are negligible, which simplifies the analysis of the diffusion equations considerably. A stochastic differential of order $\mathrm{d}t$ whose mean is zero has variance of order $(\mathrm{d}t)^2$ and therefore this is also negligible. In the Itô calculus all such negligible differentials are set to zero. The variance of $\left(\mathrm{M}\,(\mathrm{d}w)^2\right) - (\mathrm{d}w)^2$ is of order $(\mathrm{d}t)^2$, which can be neglected, so if we want to save notation, we can write

$$(\mathrm{d}w)^2 = \mathrm{d}t, \tag{2.13}$$

without the mean that appears in equation (2.11).

Following the rules of the Itô calculus, the differential of a function f of a stochastic variable x and a product of two functions f and g are

$$\mathrm{d}f = \mathrm{d}x \cdot f' + \tfrac{1}{2}(\mathrm{d}x)^2 \cdot f'', \qquad \mathrm{d}(fg) = f\mathrm{d}g + \mathrm{d}f \cdot g + \mathrm{d}f \cdot \mathrm{d}g. \tag{2.14}$$

The ordering of the terms has been retained on the right side of the second equation, because QSD sometimes has noncommuting fluctuations.

2.4 Using Itô calculus

The increase with time of the variance of the height of a Brownian particle without drift is a simple example of the Itô calculus. Suppose for simplicity that at $t = 0$ the displacement is $x = 0$. Then the variance at some later time is $\Sigma^2(x) = \mathrm{M}\,x^2$. The change in the variance after time $\mathrm{d}t$ is

$$\begin{aligned} \mathrm{d}\Sigma^2(x) &= \mathrm{d}\,\mathrm{M}\,x^2 \\ &= \mathrm{M}\,\mathrm{d}(x^2) \qquad \text{(change of the mean is mean of the change).} \end{aligned} \tag{2.15}$$

Using equation (2.14) for the differential of a product,

$$\mathrm{d}\Sigma^2(x) = \mathrm{M}\,2x\mathrm{d}x + \mathrm{M}\,(\mathrm{d}x)^2. \tag{2.16}$$

The initial displacement x is independent of the following fluctuation, so x can be taken outside the mean, and

$$\left.\begin{aligned} \mathrm{M}\,x\mathrm{d}x &= x\,\mathrm{M}\,\mathrm{d}x = xa\,\mathrm{M}\,\mathrm{d}w = 0, \\ \mathrm{d}\Sigma^2(x) &= \mathrm{M}\,(a\mathrm{d}w)^2 = a^2\mathrm{d}t, \end{aligned}\right\} \tag{2.17}$$

where the important term is the square of a stochastic differential, which cannot be neglected in the Itô calculus.

The averaged picture of Fokker and Planck does not represent the motion of the individual particles directly, but uses the probability distribution $\rho(x, t)$ of an ensemble of particles. If there is no drift, its differential equation follows from the stochastic differential equation (2.12), using the Itô calculus. For a particular fluctuation $\mathrm{d}w$, the distribution at time $t + \mathrm{d}t$ is given by shifting $\rho(x, t)$ by $\mathrm{d}x$, giving

$$\begin{aligned} \rho(x, t + \mathrm{d}t) &= \rho(x - \mathrm{d}x, t) = \rho(x - a\mathrm{d}w, t) \\ &= \rho(x, t) - a\rho'(x, t)\mathrm{d}w + \tfrac{1}{2}a^2\rho''(x, t)(\mathrm{d}w)^2, \end{aligned} \tag{2.18}$$

where the prime means differentiation with respect to x. For the whole ensemble we take the mean over all possible fluctuations. Now $\mathrm{M}\,\mathrm{d}w = 0$, $\mathrm{M}\,(\mathrm{d}w)^2 = \mathrm{d}t$ and

$$\begin{aligned} \rho(x, t + \mathrm{d}t) &= \mathrm{M}\,\rho(x + \mathrm{d}x, t) \\ &= \rho(x, t) + \tfrac{1}{2}a^2\rho''(x, t)\mathrm{d}t, \\ \text{so} \qquad \mathrm{d}\rho &= \tfrac{1}{2}a^2\rho''(x, t)\mathrm{d}t \qquad \text{(Fokker-Planck)}. \end{aligned} \tag{2.19}$$

The solution of (2.19) is the set of Gaussians illustrated in figure 2.2.

These two derivations are typical of many that appear in QSD. Because the ordinary differential calculus is used so much, most people find it difficult to avoid neglecting some of the squared differentials at first. This problem is overcome in the alternative but equivalent Stratonovich calculus [54], at the price of a slightly more difficult interpretation of the results, but the Itô form is used here.

2.5 Drift and the interval τ

When the density of the particles is greater than their fluid environment, the Earth's gravitational field adds a drift velocity v to the vertical diffusion, giving the stochastic differential equation

$$\mathrm{d}x = v\,\mathrm{d}t + a\,\mathrm{d}w, \tag{2.20}$$

where the coefficient of $\mathrm{d}t$ gives the drift velocity and the coefficient of $\mathrm{d}w$ the magnitude of the diffusion. If x is replaced by $x - \mathrm{M}\,x$, equation (2.20)

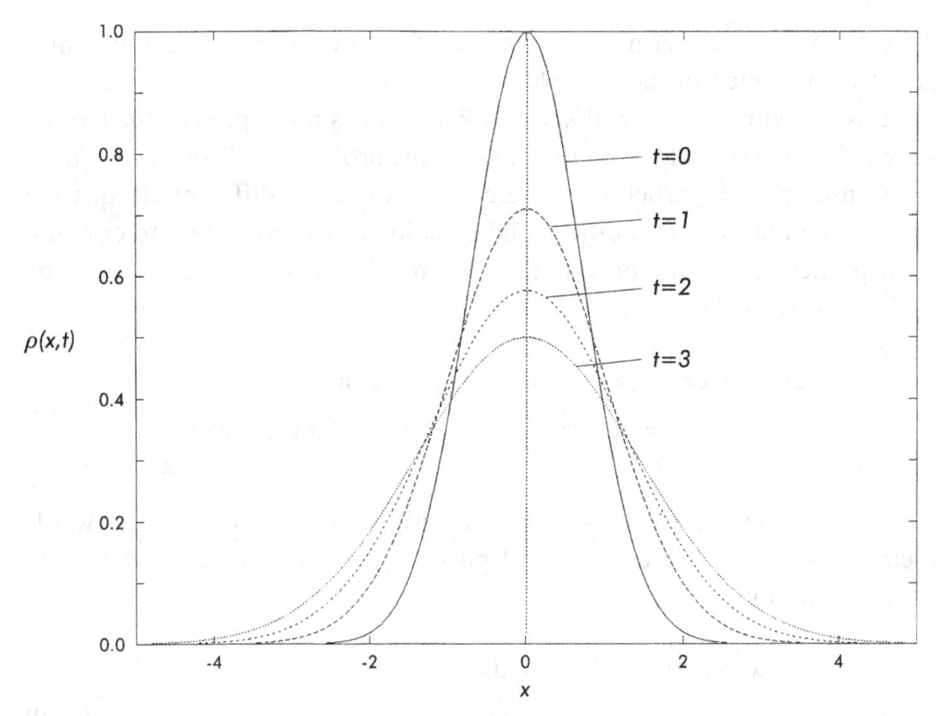

Figure 2.2 Gaussian solutions of the Fokker-Planck equation for Brownian motion without drift.

reduces to equation (2.12), so the variance of the displacement x is unaffected by the drift.

Also for more general differential stochastic processes, in which x represents the state of some dynamical system, the drift is the term containing $\mathrm{d}t$ and the diffusion terms contain fluctuations like $\mathrm{d}w$. For sufficiently small times, the diffusion, which is proportional to $(\mathrm{d}t)^{1/2}$, dominates the drift, which is proportional to $\mathrm{d}t$. The approximate boundary between the smaller time intervals dominated by the diffusion and the longer time intervals dominated by drift is obtained by supposing that equation (2.20) remains valid for finite time intervals, and equating the increments due to fluctuation and drift, giving

$$v\tau = a\tau^{1/2}, \qquad \tau = (a/v)^2 \qquad \text{(boundary interval).} \tag{2.21}$$

This is the boundary interval illustrated in figure 2.1 on page 10.

2.6 Correlation and ensemble covariance

This section follows from section 2.2 on probabilities. In any ensemble of systems, two dynamical variables f and g may or may not be correlated. If they are uncorrelated then their joint distribution is the product

$$\mathbf{Pr}(f, g) = \mathbf{Pr}(f)\mathbf{Pr}(g) \tag{2.22}$$

and a knowledge of the value of f then gives no information about the value of g.

The *ensemble covariance* of f and g,

$$\Sigma(f, g) = \mathrm{M}\,(f - \mathrm{M}f)\,(g - \mathrm{M}\,g) = \mathrm{M}\,(fg) - (\mathrm{M}f)\,(\mathrm{M}\,g) \tag{2.23}$$

is defined in terms of the deviation of f and g from their means. It is a bilinear measure of the correlation between f and g. It is zero if f and g are uncorrelated. If it is nonzero then they are correlated. But f and g can be correlated nonlinearly, in which case the covariance may be zero, despite the correlation. An example of this is when the variables x and y are distributed uniformly on a circle in the (x, y)-plane, represented by

$$x = \cos\theta, \qquad y = \sin\theta, \tag{2.24}$$

where θ is distributed uniformly between $-\pi$ and π. The variables x and y are then strongly correlated, but their covariance is zero. However, if the covariance of every function of f and every function of g is zero, then f and g are uncorrelated.

The covariance of f with itself is just the variance of f.

$$\Sigma(f,f) = \Sigma^2(f). \tag{2.25}$$

We often need to deal with the statistical properties of complex variables f, g. There is a natural definition of the variance, which is

$$\begin{aligned}\Sigma(f,f) &= \Sigma^2(f) = \mathrm{M}\,(f^*f) - (\mathrm{M}f)^*(\mathrm{M}f) = \mathrm{M}\,|f|^2 - |\mathrm{M}f|^2 \\ &= \Sigma^2(f_\mathrm{R}) + \Sigma^2(f_\mathrm{I}),\end{aligned} \tag{2.26}$$

where z_R and z_I are the real and imaginary components of a complex quantity z.

For general probability distributions in complex f and g, the properties of the covariance are not so simple, but we will not be concerned with general distributions, only with those distributions where multiplication of the important variables by a phase factor u leaves all probabilities, including joint probabilities, unchanged, so that

$$\mathbf{Pr}(uf, ug) = \mathbf{Pr}(f, g) \qquad (|u|^2 = 1). \tag{2.27}$$

In that case there is a natural generalization of the definition of ensemble variance to ensemble covariance, given by

$$\Sigma(f, g) = \mathrm{M}\,(f^* g) - (\mathrm{M} f)^*(\mathrm{M}\, g) = \Sigma(f_\mathrm{R}, g_\mathrm{R}) + \Sigma(f_\mathrm{I}, g_\mathrm{I}), \tag{2.28}$$

which has the same basic properties as the ensemble covariance of real variables. Correlation between systems is particularly important. If A and B are two systems, and AB is the combined system, then A and B are uncorrelated in an ensemble if the state x of A is uncorrelated with the state y of B, so

$$\mathbf{Pr}_\mathrm{AB}(x, y) = \mathbf{Pr}_\mathrm{A}(x)\mathbf{Pr}_\mathrm{B}(y). \tag{2.29}$$

It follows that if the covariance of every function of the state of A with every function of the state of B is zero, the systems are uncorrelated, and conversely.

Classical probabilities, correlations, variances and covariances are properties of classical ensembles, not of individual systems, even when the ensemble is not mentioned explicitly. We will see in later chapters that this is not true of quantum variances and covariances.

Two systems that have not interacted either directly or indirectly are always uncorrelated. When they do interact, they almost always become correlated. We often want to study the behaviour of one of the systems on its own, even when it is continually interacting with another system. For example, an open system is continually interacting with its environment, and the environment is usually so big and complicated that it cannot be studied in detail. Brownian motion is an example. In such cases we are only interested in the *reduced* probability distribution for the system itself. Let A be the system and B the environment. Write Tr_B for the integral over all the states y of the environment. Then the reduced distribution for A is

$$\mathbf{Pr}_\mathrm{A}(x) = \mathrm{Tr}_\mathrm{B}\mathbf{Pr}_\mathrm{AB}(x, y). \tag{2.30}$$

If we know something about the distribution of B, for example that it is in thermal equilibrium at a fixed temperature, and we know the dynamics of the

interaction between A and B, then we can often derive stochastic equations of motion for A, and deterministic evolution equations for the probability distribution $\mathbf{Pr}_A(x)$.

The dynamics of quantum mechanical amplitudes for pure quantum states has some features in common with the dynamics of classical probabilities for ensembles of classical systems. Similarly quantum entanglement, described in section 3.3, resembles classical correlation in some respects. Despite these common features, quantum amplitudes and entanglement have some special properties of their own.

2.7 Complex diffusion

For applications to quantum mechanics, isotropic diffusion in two real dimensions is best considered as one-dimensional complex diffusion in the complex z-plane illustrated in figure 2.3. The standard normalized complex stochastic differential is

$$d\xi = d\xi_R + i d\xi_I, \tag{2.31}$$

where the real and imaginary parts satisfy

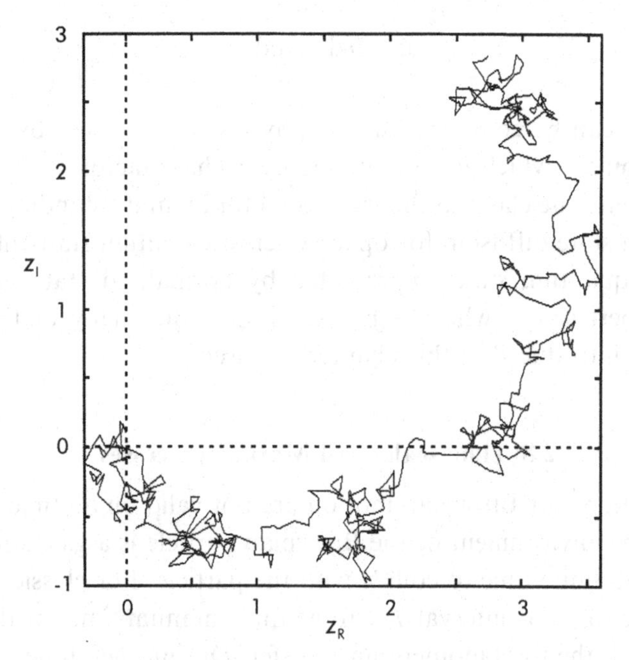

Figure 2.3 Isotropic diffusion in the complex z-plane.

$$\mathrm{M}\,\mathrm{d}\xi_{\mathrm{I}} = \mathrm{M}\,\mathrm{d}\xi_{\mathrm{R}} = 0 \qquad \text{(zero mean)},$$
$$\mathrm{M}\,\mathrm{d}\xi_{\mathrm{I}}\mathrm{d}\xi_{\mathrm{R}} = 0 \qquad \text{(zero covariance)},$$
$$\mathrm{M}\,\mathrm{d}\xi_{\mathrm{R}}^2 = \mathrm{M}\,\mathrm{d}\xi_{\mathrm{I}}^2 = \mathrm{d}t/2 \qquad \text{(normalized)}.$$

The equivalent complex form of the conditions is more convenient:

$$\begin{aligned} \mathrm{M}\,\mathrm{d}\xi &= 0 && \text{(zero mean)},\\ (\mathrm{d}\xi)^2 = \Sigma(\mathrm{d}\xi^*, \mathrm{d}\xi) &= 0 && \text{(zero covariance)},\\ |\mathrm{d}\xi|^2 = \Sigma^2(\mathrm{d}\xi) &= \mathrm{d}t && \text{(normalized)}, \end{aligned} \tag{2.33}$$

where because of the isotropy the statistics of the motion is unaffected if $\mathrm{d}\xi$ is multiplied by a phase factor of unit magnitude.

The equations (2.14) for the differential of a product and of a function apply to complex fluctuations as they do to real ones, but there is an important simplification for complex fluctuations, because the square of a complex differential fluctuation is zero, so the second term in the expression for $\mathrm{d}f$ vanishes. This leads to a simpler quantum state diffusion theory when complex fluctuations are used.

The Langevin-Itô equation of a particle with drift and diffusion at position $z(t)$ in the complex plane is

$$\mathrm{d}z = v\mathrm{d}t + a\mathrm{d}\xi. \tag{2.34}$$

Both v and a can be complex, but the physics is unaffected by the (gauge) transformation for which a is multiplied by a phase factor.

This completes the classical theory needed for an understanding of elementary quantum state diffusion for open systems. Quantum state diffusion is a diffusion of quantum states represented by normalized state vectors in a complex Hilbert space, which is not so simple. But many of the methods and notation introduced in this chapter are used.

2.8 Time scales and Markov processes

The Itô equations for Brownian motion are not valid on all time scales. For example, if the environment of the Brownian particle is a gas, a molecule of the gas takes a finite time to collide with the particle. For classical collisions we could take the time interval T_{col} to be the minimum time for the collision to produce half the total momentum transfer. During such time intervals the

motion looks smooth and cannot be represented by an Itô equation like (2.12) or (2.20). There are correlations between the fluctuations over time intervals comparable to T_{col} and smaller, so that on this scale the Markov assumption that there are no correlations is wrong. But for time scales that are long by comparison, the correlations are negligible and the Itô equations can be used. Typically T_{col} is smaller than the boundary time τ by many orders of magnitude, and cannot be detected experimentally.

In the theory of Brownian motion, a uniform drift has been assumed, which seems incompatible with Newton's second law of motion. A heavy body in air, like a skydiver's, first accelerates. The resistance of the air then reduces her acceleration towards zero, so that she approaches a terminal drift velocity. Her Brownian fluctuations are negligible. For Brownian particles the period of acceleration is normally so short that it can be neglected. Only the terminal drift velocity and the fluctuations can be seen.

For dense gases and liquids the time between collisions is very short compared with the collision time. However, for diffusion in gases of very low density, it may be very much longer, and when the mean free path is macroscopic, the Markov assumption and Itô equations may not be valid on any scale. The quantum mechanics of the collisions complicates things further. It was Einstein's achievement to recognize that the relevant theory was usually independent of such complications. That is how he produced his theory more than two decades before any proper understanding of the elementary collisions which cause the fluctuations.

Similar considerations apply to quantum state diffusion. The Markovian Itô theory of quantum state diffusion is valid on time scales that are long compared with the basic time scale for interaction with the environment, but sometimes the correlations between successive fluctuations cannot be ignored, the evolution of the quantum system is non-Markovian, and QSD cannot be used.

2.9 Many fluctuations

This section is a little bit more difficult than the rest of the chapter. It includes the classical theory needed for all but the simplest applications of QSD, and can be skipped at first reading. Definitions for single fluctuations are generalized to many fluctuations, which introduces some mathematical complications.

Suppose there are K real fluctuations $\mathrm{d}w'_k$ with zero mean. We cannot always rule out the possibility that the fluctuations $\mathrm{d}w'_k$ are correlated, or even linearly dependent. Since powers of $\mathrm{d}w'_k$ higher than squares are negligible, the means and variances and covariances provide all the statistical information about them, so

$$\left.\begin{aligned} \mathrm{M}\,\mathrm{d}w'_k &= 0 && \text{(zero mean)}, \\ \Sigma(\mathrm{d}w'_k, \mathrm{d}w'_{k'}) = \mathrm{M}\,\mathrm{d}w'_k \cdot \mathrm{d}w'_{k'} &= X'_{kk'}\,\mathrm{d}t && \text{(general)}, \end{aligned}\right\} \tag{2.35}$$

thus defining $X'_{kk'}$.

The variance of a fluctuation is treated as a norm, and the covariance of two fluctuations as a scalar product. By taking linear combinations of the fluctuations we can transform them to a standard form, in which they are all normalized to $\mathrm{d}t$ and orthogonal. This makes an orthonormal set of fluctuations. Normalization is no problem, because fluctuations are always used with coefficients, and unwanted constant factors can be absorbed into the coefficients. Orthogonality is the problem.

The symmetric matrix $X'_{kk'}$ can be diagonalized by an orthogonal transformation:

$$X_{jj'} = \sum_{kk'} O_{jk} X'_{kk'} O^{\mathrm{tr}}_{k'j'}, \tag{2.36}$$

giving a new set of fluctuations that are linear combinations of the old ones:

$$\mathrm{d}w^0_j = \sum_k O_{jk} \mathrm{d}w'_k. \tag{2.37}$$

For the new fluctuations, the off-diagonal covariances are zero and, if some of the old fluctuations $\mathrm{d}w'_k$ are linearly dependent, some of the new variances $\mathrm{d}w^0_j$ are zero too. The latter may be ignored, as they do not contribute to the diffusion. Suppose there are J remaining. Generally they are not normalized, but they can be normalized so that each variance is $\mathrm{d}t$, giving a set of orthonormal fluctuations:

$$\begin{aligned} \mathrm{M}\,\mathrm{d}w_j &= 0 && \text{(zero mean)}, \\ \Sigma(\mathrm{d}w_j, \mathrm{d}w_{j'}) = \mathrm{M}\,\mathrm{d}w_j \cdot \mathrm{d}w_{j'} &= \delta_{jj'}\mathrm{d}t && \text{(orthonormality)}. \end{aligned} \tag{2.38}$$

This is the standard normalization, in which the norm of the vector fluctuation $\{\mathrm{d}w_j\}$ is $J\mathrm{d}t$.

There is a direct generalization to diffusion in a complex vector space, which is needed for QSD. A general diffusion has correlated fluctuations $\mathrm{d}\xi'_k$ and nonzero off-diagonal covariances, using the definition of complex covariance of section 2.6 in equation (2.28). The Hermitian matrix $X'_{kk'}$ can be diagonalized by a unitary change of basis. For the new fluctuations $\mathrm{d}\xi_k^0$, the off-diagonal covariances are zero and, if some of the fluctuations $\mathrm{d}\xi'_k$ are linearly dependent, some of the diagonal variances of $\mathrm{d}\xi_j^0$ are zero too, and there are only $J < K$ independent fluctuations. These can be normalized.

The new complex fluctuations have the standard form

$$\begin{aligned} \mathrm{M}\,\mathrm{d}\xi_j &= 0 && \text{(zero mean)}, \\ \Sigma(\mathrm{d}\xi_j, \mathrm{d}\xi_{j'}) &= \mathrm{M}\,\mathrm{d}\xi_j^* \cdot \mathrm{d}\xi_{j'} = \delta_{jj'}\mathrm{d}t && \text{(orthonormality)}. \end{aligned} \tag{2.39}$$

The standard forms are not unique. Orthogonal transformations of $\{\mathrm{d}w_j\}$ and unitary transformations of $\{\mathrm{d}\xi_j\}$ leave their properties are unchanged. These may be considered as different fluctuations or as a different representation of the same vector fluctuation in the linear fluctuation space.

A general Itô equation for a real variable x with many independent real fluctuations is

$$\mathrm{d}x = v\mathrm{d}t + \sum_j a_j \mathrm{d}w_j. \tag{2.40}$$

Let $f(x, t)$ be a differentiable function of x and of t. Because it is differentiable with respect to t it does not depend explicitly on the fluctuations. The change in $f(x, t)$ in a time $\mathrm{d}t$ is

$$\mathrm{d}f(x, t) = (\partial f/\partial x)\mathrm{d}x = (\partial f/\partial x)v\mathrm{d}t + \sum_j (\partial f/\partial x)a_j \mathrm{d}w_j. \tag{2.41}$$

Consequently the contributions to $\mathrm{d}f$ from the drift term and each of the fluctuation terms proportional to $\mathrm{d}w_j$ are additive. This is the *additivity rule*, which is also true for complex x with complex fluctuations, when x is a vector, or even when it is a quantum state vector, as for QSD. The additivity rule is used extensively in QSD to obtain the theory of complicated systems from the theory of elementary systems.

3

Open quantum systems

An individual open quantum system can be represented by a pure state, and the statistical properties of an ensemble of open systems by a probability distribution over pure states. The density operator which can be obtained from this distribution is more compact, but does not represent the dynamics of the individual systems of the ensemble. A density operator is also used to represent an open quantum system that is entangled with its environment as the result of interaction. Measurement and preparation are given general definitions in terms of the interaction of a quantum system with its environment. The boundary between system and environment is ambiguous. For a pure state there are quantum expectations, quantum variances and quantum covariances.

3.1 States of quantum systems

A pure state of a quantum system with a state vector $|\psi\rangle$ that satisfies the Schrödinger equation is a very useful fiction.

Even if the system is isolated, every system we know has interacted with other systems in the past, and so has become entangled with them, as described in section 3.3. Strictly speaking, no system is independent, and so no system can be properly represented by a state vector $|\psi\rangle$. All systems are open, with the exception of the whole universe. However, quantum systems are commonly represented by state vectors, theories based on this representation are very successful, and we use this representation too.

The representation of the same pure state by a projection operator or *projector*

$$\mathbf{P} = \mathbf{P}_\psi = |\psi\rangle\langle\psi|, \tag{3.1}$$

is equivalent to state vector representation, except for a physically irrelevant external phase factor. We use both.

The Schrödinger time evolution of a pure quantum state $|\psi\rangle$ with Hamiltonian $\mathbf{H}$ is given by the differential equations

$$|\dot{\psi}(t)\rangle = -(i/\hbar)\mathbf{H}|\psi(t)\rangle \qquad \text{and} \qquad \dot{\mathbf{P}} = -(i/\hbar)[\mathbf{H}, \mathbf{P}], \tag{3.2}$$

from which it follows that the evolution over finite times is given by

$$|\psi(t)\rangle = \exp(-i\mathbf{H}t/\hbar)|\psi(0)\rangle \quad \text{and} \quad \mathbf{P}(t) = \exp(-i\mathbf{H}t/\hbar)\mathbf{P}(0)\exp(i\mathbf{H}t/\hbar). \tag{3.3}$$

This evolution is linear. It is also *deterministic*, meaning that the present state determines future states uniquely. The norm $\langle\psi|\psi\rangle$ of the state vector and the corresponding trace $\mathrm{Tr}\,\mathbf{P}$ of the projector are preserved. Linear norm-preserving transformations are unitary. Quantum state diffusion preserves the norm and the trace, but it is neither linear nor deterministic. It is nonlinear and stochastic.

An open quantum system continually interacts with its environment. An ensemble of these systems is sometimes called a 'mixed state' of the system and is represented by a density operator which is defined in the next section, 3.2. But in QSD we usually use the representation of a single state of the ensemble by a pure state $|\psi\rangle$. For open quantum systems, just like isolated ones, this is a very useful fiction. We also use the density operator, but it plays a secondary role.

The properties of systems with many orthogonal states are easier to understand when there are only two, as in the example of a spin one-half system. This chapter uses these 'two-state' systems to illustrate mathematically and pictorially the properties of the more general systems that are needed for QSD.

By convention, the states $|+\rangle$ and $|-\rangle$ with 'up' and 'down' spins in the $+z$ and $-z$ direction are used as the base vectors $(1, 0)$ and $(0, 1)$ respectively. These are the eigenvectors of the Pauli matrix $\boldsymbol{\sigma}_z$. The eigenvectors of $\boldsymbol{\sigma}_x$ represent the spins in the $\pm x$ direction and of $\boldsymbol{\sigma}_y$ represent the spins in the $\pm y$ directions. The matrix representations of the operators and eigenvectors can be found in most quantum mechanics texts.

The states and ensembles of 'two-state' systems are best visualized in the three-dimensional linear space spanned by the projectors, which is *not* the same as the Hilbert space of the states themselves. The projectors are points on the surface of the Bloch or Poincaré sphere, illustrated in figure 3.1. The projectors $\mathbf{P}$ can be represented by 2×2 matrices. In the space of the projectors, the centre of the sphere, which is the origin of the coordinates, is at half the unit matrix, $\frac{1}{2}\mathbf{I}$. The vector representing the projector $\mathbf{P}$ is

$$\vec{r} = (x, y, z) = \mathrm{Tr}\mathbf{P}\vec{\sigma}, \qquad \text{where} \qquad \vec{\sigma} = (\sigma_x, \sigma_y, \sigma_z) \tag{3.4}$$

and the projector corresponding to the vector $\vec{r}$ is

$$\mathbf{P} = \tfrac{1}{2}(\mathbf{I} + \vec{r} \cdot \vec{\sigma}). \tag{3.5}$$

The projectors of the two eigenstates of a general spin-half matrix $\boldsymbol{\sigma}$ are at $\frac{1}{2}(\mathbf{I} \pm \boldsymbol{\sigma})$, which are on opposite sides of the sphere. A point on the surface of the sphere indicates the direction of the spin in three-dimensional position space. The up and down spins are at the north and south poles.

Besides spin-half systems, other examples of two-state systems are polarization states of photons, the two states of an atom that resonate strongly with laser light of a single fixed frequency, and the two occupation states for a fermion. For photons, the up and down spins correspond to vertical and

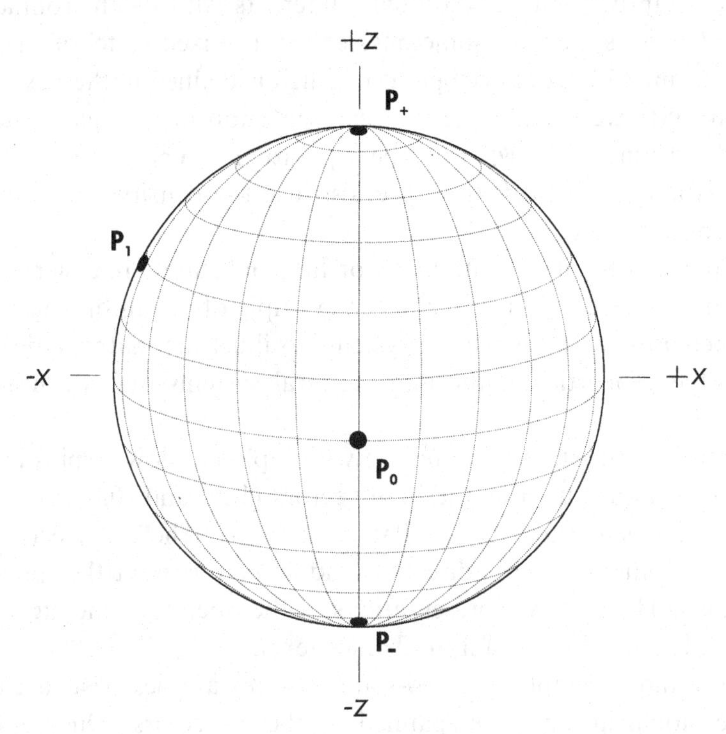

Figure 3.1 Projectors as points on the surface of the Bloch sphere.

horizontal polarizations of the photon, the eigenstates of $\boldsymbol{\sigma}_x$ to linear polarization at 45° and the eigenstates of $\boldsymbol{\sigma}_y$ to circular polarizations. For two-state atoms, the up and down spins correspond to the excited and ground states of the atom and the other spins to coherent combinations of the two atomic states.

Two-state quantum systems, like two-state classical systems, can be used to represent binary numbers. The corresponding bits are called *qubits*, spoken as 'kewbits'. Quantum systems of many qubits behave very differently from classical systems with many classical bits. For example, simultaneous parallel operations can be performed on qubits that are impossible with classical bits. This is the basis of modern quantum technology, including quantum communication, quantum cryptography and possible future quantum computers. QSD has been applied to quantum computers, which are discussed further in chapter 6.

All two-state Hamiltonians have the form $\hbar\omega\boldsymbol{\sigma}$, where ω is an angular frequency and $\boldsymbol{\sigma}$ is a spin operator in an arbitrary direction. The Schrödinger evolution of a pure quantum state is then given by uniform rotation on the surface of the Bloch sphere with angular velocity ω about an axis through the eigenstates of $\boldsymbol{\sigma}$. For $\mathbf{H} = \omega\boldsymbol{\sigma}_z$, a state moves uniformly around a line of latitude.

The book by Allen and Eberly [2] has a useful introduction to the Bloch sphere.

3.2 Ensembles of quantum systems

Classical dynamics and Schrödinger dynamics are both deterministic, but there is an unavoidable stochastic behaviour where they interact. Because of this, probability is more important for quantum theory than it is for classical theories.

A single system whose state is not perfectly known, or whose evolution is stochastic, is represented by an ensemble of systems with a probability distribution over pure states. One example is an open quantum system, like a molecule in a fluid. Another is a collection of many quantum systems, like many runs of an experiment.

Depending on the type of ensemble, the probabilities can be represented by a continuous probability distribution $\mathbf{Pr}(\psi)$ in the very large space of all quantum states $|\psi\rangle$ with projectors $\mathbf{P}_\psi$, or a discrete distribution, which consists of a set of probabilities $\mathbf{Pr}(j)$ for states $|j\rangle$ with projectors $\mathbf{P}_j$. The states need not be orthogonal in either case. For simplicity, the theory is developed here for discrete distributions, which can be generalized without difficulty to continuous distributions, replacing sums by integrals where appropriate. Continuous distributions are discussed in [6].

For quantum state diffusion theory, the probability distribution is the most important representation of an ensemble of quantum systems. But it is not a common representation, because the measurable properties are all given by a simpler representation using density operators.

The *density operator* and the corresponding vector for an ensemble of quantum systems are

$$\rho = \sum_j \mathbf{Pr}(j)\mathbf{P}_j, \qquad \vec{r} = \mathrm{Tr}\rho\vec{\sigma} \tag{3.6}$$

and the point $\vec{r}$ is in the interior of the Bloch sphere unless the members of the ensemble are identical.

Remarkably, ensembles with the same density operator cannot be distinguished from one another by measurements, however they may be performed, so that for many purposes such ensembles are equivalent. This equivalence is a major theme of this book. However, a density operator does not normally represent the individual systems of the ensemble. An exception is the very special type of ensemble which consists of identical pure states, for which the ensemble is represented by the pure state projector for that state.

The probability of the system being found by a measurement in the state $|\psi\rangle$ is

$$\mathrm{Tr}\rho\mathbf{P}_\psi = \langle\psi|\rho|\psi\rangle. \tag{3.7}$$

This property is sometimes used to define the density operator. The probabilities must be positive or zero, and their sum for a complete set of orthonormal states is unity. In other words, the diagonal elements of any matrix representation of ρ must be non-negative and sum to unity. These two conditions are the positivity and trace conditions on the density operator.

An ensemble of quantum systems may start as an ensemble of identical pure states and then evolve into an ensemble of differing pure states, just as an ensemble of Brownian particles starting at one place evolves into an ensemble at different places. Examples are quantum Brownian motion, radioactive decay, optical transitions in atoms and nuclei, and quantum measurements. All these processes can be treated using the theory of open quantum systems.

An ensemble of different pure states of a two-state system can be represented by a probability distribution on the surface of the Bloch sphere, or by a point in its interior representing the density operator. The point representing the density operator is a linear combination of the points on the surface, with weights given by their probabilities, like a centre of mass. If there are only two pure states in the ensemble, then the density operator lies on the line joining their projectors. The inverse process of *unravelling* a density operator into its component pure states is not unique, because the same density operator can be formed from many different combinations of pure states, as illustrated in figure 3.2, unless the ensemble consists of many copies of a single pure state.

The Schrödinger dynamics of a probability distribution $\mathbf{Pr}(\psi)$ over pure states is given directly by the evolution of the individual pure states of the ensemble. For two-state systems, the distribution on the surface of the sphere is rotated about the axis determined by the Hamiltonian, as illustrated in figure 3.3.

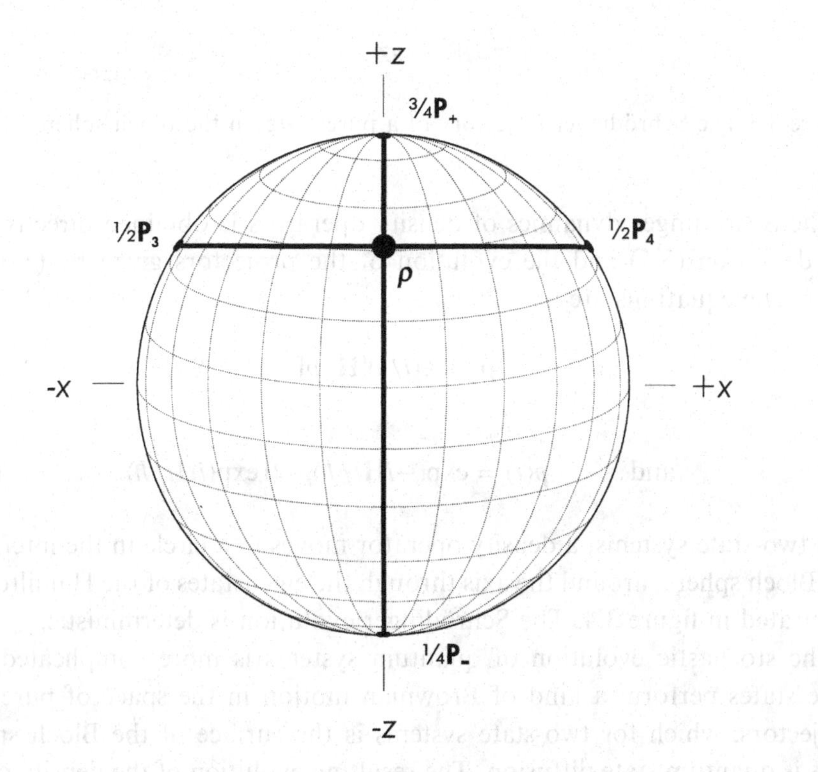

Figure 3.2 Two unravellings of a density operator into pure state projectors. The density operator $\boldsymbol{\rho}$ on the axis joining the north and south poles is a linear combination of the pure state projectors at the end of the same straight line.

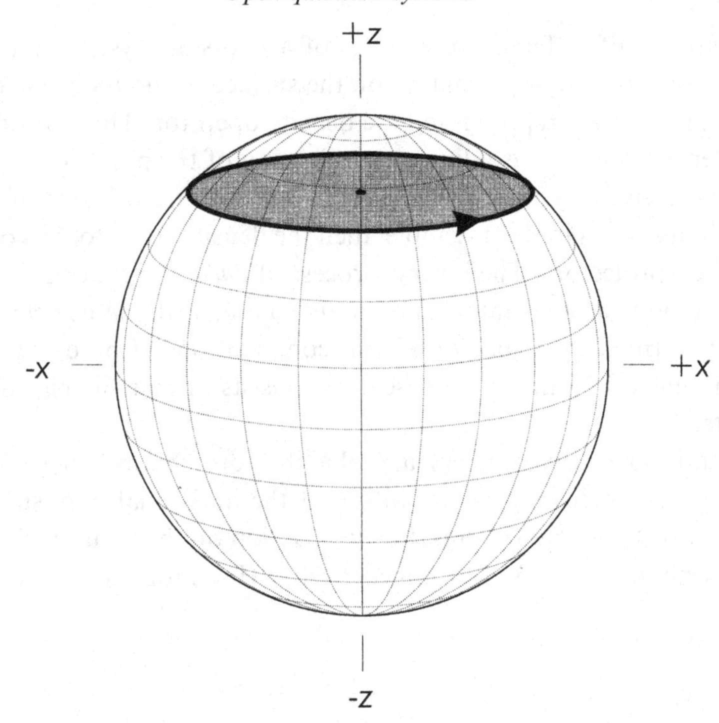

Figure 3.3 The Schrödinger trajectory of a pure state on the Bloch sphere.

The Schrödinger dynamics of density operators is obtained directly from the definition (3.6) and the evolution of the projectors given by (3.2) and (3.3). The equations are

$$\dot{\rho} = -(i/\hbar)[\mathbf{H}, \rho] \tag{3.8a}$$

and

$$\rho(t) = \exp(-i\mathbf{H}t/\hbar)\rho(0)\exp(i\mathbf{H}t/\hbar). \tag{3.8b}$$

For two-state systems, a density operator moves in a circle in the interior of the Bloch sphere, around the axis through the eigenstates of the Hamiltonian, illustrated in figure 3.4. The Schrödinger evolution is deterministic.

The stochastic evolution of quantum systems is more complicated. The pure states perform a kind of Brownian motion in the space of pure state projectors, which for two-state systems is the surface of the Bloch sphere. This is quantum state diffusion. The resulting evolution of the density operator, like the classical probability density of the previous chapter, is deter-

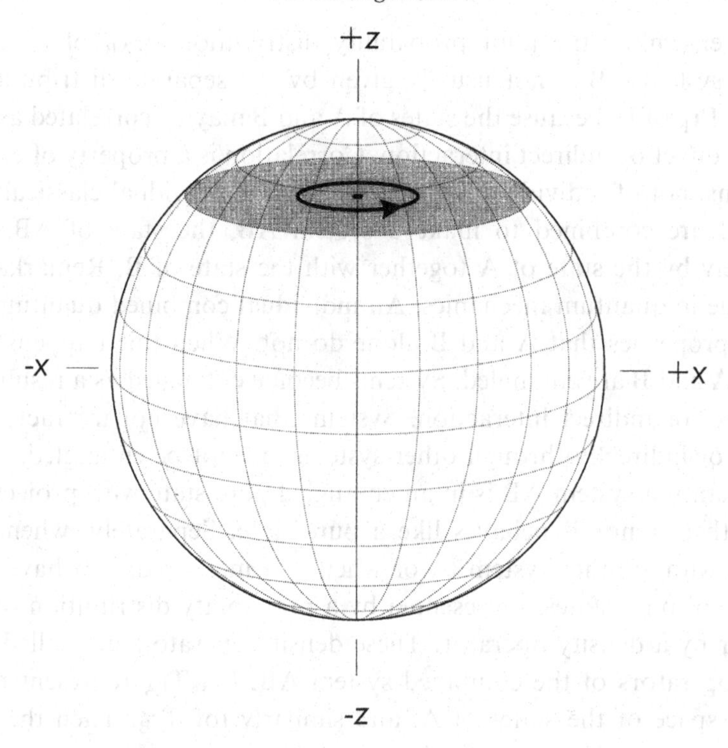

Figure 3.4 The Schrödinger trajectory of a density operator in the interior of the Bloch sphere.

ministic, and satisfies a linear differential equation known as the master equation. These processes are the subject of the following chapters.

3.3 Entanglement

Entanglement is one of the most puzzling properties of quantum systems. The concept was introduced by Schrödinger in 1935 [131, 132, 133, 154]. His briefest definition of *entanglement* was 'the whole is in a definite state whereas the parts are not'. In modern work on Bell inequalities the word is sometimes used in a narrower sense, but the original broad sense will be used here. For open systems, the system together with its environment is in a definite pure quantum state, but the system alone is not.

Einstein himself had difficulty with entanglement, perhaps to the point of disbelief [47, 154]. But it is important to understand it, because it lies behind the quantum theory of open systems and also many of the most fascinating recent quantum experiments, including tests of the Bell inequalities, quantum cryptography, quantum teleportation and proposals for quantum computers.

In an ensemble, the joint probability distribution $\mathbf{Pr}_{AB}$ of a combined classical system AB is not usually given by the separate distributions $\mathbf{Pr}_A$ of A and $\mathbf{Pr}_B$ of B, because the states of A and B may be correlated as a result of a past direct or indirect interaction. Correlation is a property of ensembles of systems, not of individual systems. But if two individual classical systems A and B are combined to make a system AB, the state of AB is given completely by the state of A together with the state of B. Remarkably this is not true in quantum mechanics. An individual combined quantum system AB has properties that A and B alone do not. When this happens the two systems A and B are entangled. Systems become entangled as a result of their past direct or indirect interaction. Systems that have not interacted, either directly, or indirectly through other systems, cannot be entangled.

If a quantum system AB is in an entangled pure state with projector $\mathbf{P}_{AB}$, then neither A nor B behaves like a pure state. Separately, when A or B interacts with another system C, or when it is measured, it behaves like an ensemble of pure states, represented by a probability distribution over pure states, or by a density operator. These density operators are called *reduced* density operators of the combined system AB. Let Tr_A represent the trace over the space of the states of A, and similarly for Tr_B. Then the reduced density operators for A and for B are

$$\rho_A = \mathrm{Tr}_B\, \rho_{AB} \qquad \text{and} \qquad \rho_B = \mathrm{Tr}_A\, \rho_{AB}. \tag{3.9}$$

One of the simplest examples of entanglement is for a spin-zero system that decays into two spin-half systems A and B, which then move apart. The spin states of A and B are then entangled, with pure spin state

$$|\chi_{AB}\rangle = (1/\sqrt{2})(|A+\rangle|B-\rangle - |A-\rangle|B+\rangle). \tag{3.10}$$

If we consider A as the system and B as the environment, the density operator of A is given by

$$\rho_A = \mathrm{Tr}_B \rho_{AB} = \mathrm{Tr}_B |\chi_{AB}\rangle\langle\chi_{AB}| = \tfrac{1}{2}\mathbf{I}_A, \tag{3.11}$$

where $\mathbf{I}_A$ is the identity in the spin-space of A. The density operator $\frac{1}{2}\mathbf{I}_A$ is at the centre of the Bloch sphere, representing the spherically symmetric spin distribution. A measurement of the spin of system A or system B alone gives this boring result. But simultaneous measurements of the spins of A and B in different directions tell us about the Bohm form of the Einstein-Podolsky-Rosen thought experiment, the Bell inequalities, and subsequent laboratory experiments. This is an example of entanglement between particles. The measurements show a type of correlation of the spins that cannot be produced by classical means. Entanglement in a pure state is not an actual correlation, but measurement of entangled pure states produces a correlated ensemble, so entanglement is potential correlation of a special type.

Systems are not always defined with fixed numbers of particles. Even in classical statistical mechanics the number of particles is sometimes allowed to vary, and this is essential for any quantum field theory. A system can be defined as the state of a region of space, in which there may be a varying number of particles.

An example of entanglement of systems defined in this way is quantum interference, as illustrated by the two slits experiment for individual photons shown in figure 3.5. We suppose a photon is emitted in a short pulse. The systems A and B are the contents of the regions illustrated when the pulse has just gone through the slits. At this time the system A behaves just like an ensemble with probability $\frac{1}{2}$ of containing one photon, and probability $\frac{1}{2}$ of containing zero photons. The same applies to B. A measurement of the photon number in A and in B is a measurement of the state of the combined system AB. Of the four possible number states of AB, only the two with one photon are seen, so that the measurement shows a correlation between A and B. Before the measurement, this is a potential correlation of the systems A and B, so A and B are entangled. Later on, AB shows the well-known interference properties, which are not obtained by combining the density opera-

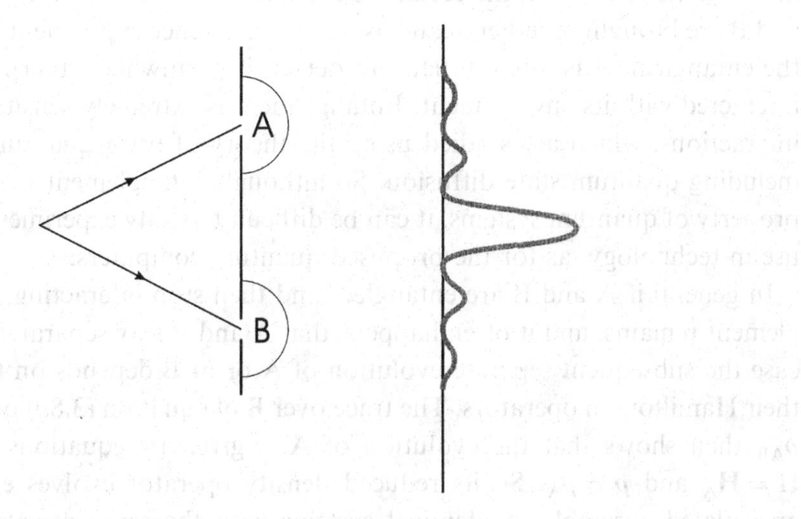

Figure 3.5 A two slits experiment with individual photons.

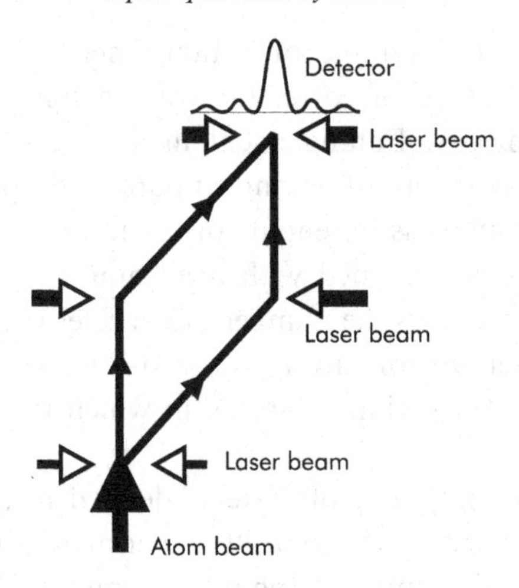

Figure 3.6 A sketch of a Mach-Zehnder atom interferometer.

tors for A and B separately. It will not show these properties if the number of photons in A or B is measured, *even* if the result of the measurement is that no photon is seen. The measurement destroys the entanglement. The same behaviour is seen in the Mach-Zehnder atom interferometer illustrated in figure 3.6. Readers are warned that some physicists who work on Bell inequalities do not use 'entanglement' for this case. Section 8.6 has more on atom interferometers.

This illustrates the general properties of entanglement, which only appear in correlations between the results of measurements on A and B, or when A and B are brought together again, as in an interference experiment. Even then the entanglement is often difficult to detect, if meanwhile either A or B has interacted with its environment. Entanglement is extremely sensitive to such interactions, which are studied using the theory of open quantum systems, including quantum state diffusion. So although entanglement is a universal property of quantum systems, it can be difficult to study experimentally or to use in technology, as for the proposed quantum computers.

In general if A and B are entangled, and then stop interacting, the entanglement remains, and it often happens that A and B stay separated, in which case the subsequent separate evolution of A or of B depends on the sum of their Hamiltonian operators. The trace over B of equation (3.8*a*) or (3.8*b*) for ρ_{AB} then shows that the evolution of A is given by equations (3.8) with $\mathbf{H} = \mathbf{H}_A$ and $\rho = \rho_A$. So its reduced density operator evolves exactly like an isolated ensemble of identical systems with the same density operator ρ_A, and experiments on A alone cannot distinguish the two. The same applies to B. But simultaneous measurements on A and B can give results that are completely different from measurements on the two isolated and unentangled ensembles. This is the key to quantum nonlocality properties in general and experiments on Bell's inequalities in particular.

3.4 Open systems

An open classical system is a system that interacts with its environment, thus producing correlation between them. An *open quantum system* A is defined here as a system that interacts with its environment B, producing entanglement of A and B. But the theory of open systems is mainly concerned with the properties of the system A. These depend on past interactions with the environment, which produce entanglement, but entanglement as such is not usually a major concern of the theory. Typically the environment is much more complicated than the system, and it is impractical to treat the combined system AB. An example is an atom in a gas or in a radiation field, like an atom near the surface of a star. Another is the radiation field of a gas laser, whose environment is a gas of atoms with inverted populations. Another is a resistor in a quantum circuit, in which the environment consists of the internal freedoms of the resistor. Another is a protein in the messy environment of a bacterium. Only in some of these systems is the environment B spatially separated from the system A.

In some cases the environment remains in approximate thermal equilibrium at a constant temperature, when it is often called a heat bath. In other cases, as for example the interaction of atoms with a radiation field, the combined system AB may be divided in different ways into system and environment. The internal states of an atom may be the system, and the radiation field may be the environment, as in matter interferometry. Or a few modes of the radiation field may be the system and the internal states of the atoms may be the environment, as for a gas laser. The system may be an atom, and the environment may be some measuring apparatus, as when a state of the atom is measured.

Let A be a system and B its environment. The outer boundary of B can be chosen so large that, as far as A is concerned, the interaction of AB with *its* environment can be neglected. Then AB can be taken to evolve by Schrödinger evolution, and the evolution of the density operator ρ_A is obtained by taking the trace:

$$\rho_A(t) = \mathrm{Tr}_B \exp(-i\mathbf{H}t/\hbar)\rho_{AB}(t)\exp(i\mathbf{H}t/\hbar). \tag{3.12}$$

Use 1 and 2 to label two ensembles of the combined system AB, with density operators ρ_{AB1} and ρ_{AB2}. At time $t = 0$ a new ensemble of systems AB is made of a sample from ensemble 1 with probability $\mathbf{Pr}(1)$ and a sample from ensemble 2 with probability $\mathbf{Pr}(2)$. Then its density operator is

$$\rho_{AB}(0) = \mathbf{Pr}(1)\rho_{AB1}(0) + \mathbf{Pr}(2)\rho_{AB2}(0). \tag{3.13}$$

By the linearity of Schrödinger evolution, the same linear combination then applies for all times t. By taking the trace over B, we have that

$$\begin{aligned} &\text{when} \quad \rho_A(0) = \mathbf{Pr}(1)\rho_{A1}(0) + \mathbf{Pr}(2)\rho_{A2}(0), \\ &\text{then} \quad \rho_A(t) = \mathbf{Pr}(1)\rho_{A1}(t) + \mathbf{Pr}(2)\rho_{A2}(t) \qquad \text{(linear evolution).} \end{aligned} \tag{3.14}$$

So, for this case, the evolution of the density operator of A is also linear, even though it does not evolve according to Schrödinger.

Almost always, in practice, the environment is so big and complicated that we cannot solve for AB in order to get the evolution of A, and so we have to approximate. As would be expected, every one of these approximations gives a linear evolution for the density operator ρ_A.

In section 7.7 we see that the linearity of density operator evolution is universal and fundamental; it does not depend on any special assumptions or approximations. By contrast, the next chapter on quantum state diffusion shows that the evolution of a state vector for an open system is usually nonlinear.

3.5 Measurement and preparation

We start with a theory of quantum measurement which is not very far from the usual one, and later develop it in stages into something very different. Measurement is a physical process by which the state of a quantum system influences the value of a classical variable. The meaning of *quantum measurement* is here extended to any such process, including laboratory measurements, but also other, very different, processes.

Laboratory measurements include the usual particle states that produce the bubbles and sparks of experimental high-energy physics, photon states that produce silver grains in photographic emulsions, and excited states of atoms that produce small currents in solid-state detectors.

Other measurements include the photons that send impulses through the optic nerves of owls, the cosmic rays that produced small but permanent dislocations in mineral crystals during the Jurassic era, and the quantum fluctuations in the early universe that are believed to have caused today's anisotropies in the universal background radiation and in galactic clusters. It includes those quantum fluctuations that are amplified by chaotic dynamics to produce significant changes in classical dynamical variables. For such measurements there is no measuring apparatus in the usual sense.

In all cases, the system that does the measuring is called the *measurer*.

A measurement takes time, because a quantum system takes time to influence a classical variable, but in the elementary quantum theory of measurement this time is ignored. If it is treated as a physical process at all, measurement is supposed to be instantaneous. Measurement that lasts a finite time appears in the next chapter. In this chapter we take it to be instantaneous.

A measurement on an ensemble of identical pure states of a quantum system typically produces a probability distribution over both the quantum states of the system and the classical states of the measurer. This distribution is conditional on the measurement being made, and so the probabilities are conditional or potential probabilities. Considering them as actual probabilities before any measurement is made leads to confusion, because in quantum mechanics there are incompatible measurements.

A measurement is an interaction between the quantum system A that is being measured and the measurer B, which can be considered as the environment of A. Other parts of the environment may interact with the system during the measurement, which experimenters try to avoid in laboratory measurements. We can assume here that these interactions are negligible. In measurement theory, as in any theory of open quantum systems, there is always a problem as to where to put the boundary between the system and its environment. This problem is discussed in section 3.6. Here the system is what is being measured and the environment is the measurer.

Let $|\psi\rangle$ be an arbitrary state of a quantum system in an ensemble of identical states with density operator $\boldsymbol{\rho} = \mathbf{P}_\psi$. Let $\mathbf{G}$ be the Hermitian operator for one of its dynamical variables, where for simplicity the spectrum of $\mathbf{G}$ is discrete and nondegenerate with eigenvalues g, eigenstates $|g\rangle$ and projectors $\mathbf{P}_g = |g\rangle\langle g|$. Unless $|\psi\rangle$ is itself an eigenstate of $\mathbf{G}$, a measurement of g produces an ensemble of different eigenstates of the quantum system, with corresponding values of some classical variable of the measuring system. The probability of making the transition to a particular state $|g\rangle$ is

$$\mathbf{Pr}(g) = |\langle g|\psi\rangle|^2 = \mathrm{Tr}\,\mathbf{P}_g\mathbf{P}_\psi. \tag{3.15}$$

During a measurement, an individual system evolves stochastically. It jumps to state $|g\rangle$ with probability $\mathbf{Pr}(g)$. The new state and projector are

$$|\psi'\rangle = |g\rangle = \langle g|\psi\rangle^{-1}\mathbf{P}_g|\psi\rangle, \tag{3.16a}$$

$$\mathbf{P}'_\psi = \mathbf{P}_g = \mathbf{Pr}(g)^{-1}\mathbf{P}_g\mathbf{P}_\psi\mathbf{P}_g. \tag{3.16b}$$

Because of the normalization factors $\langle g|\psi\rangle^{-1}$ and $\mathbf{Pr}(g)^{-1}$, this is nonlinear in $|\psi\rangle$ and nonlinear in $\mathbf{P}_\psi$, so it cannot be represented by any kind of Schrödinger equation, stochastic or otherwise. This jump process is a short time limit of the quantum state diffusion representation of measurement, as illustrated in the next chapter and demonstrated for a special case in section 5.3.

On the other hand, the new density operator obtained as a result of the measurement of the ensemble of identical pure states $|\psi\rangle$ with initial density operator $\boldsymbol{\rho} = \mathbf{P}_\psi$ is

$$\boldsymbol{\rho}' = \sum_g \mathbf{Pr}(g)\mathbf{P}_g = \sum_g \mathbf{P}_g\boldsymbol{\rho}\mathbf{P}_g, \tag{3.17}$$

which is a linear function of the original density operator $\boldsymbol{\rho}$. The final state $\boldsymbol{\rho}'$ of the density operator is obtained by taking a mean over the final projectors of (3.16), with weights given by the probabilities. The nonlinearities cancel on taking the mean. This sudden transition from $\boldsymbol{\rho}$ to $\boldsymbol{\rho}'$ is the short-time limit of a linear master equation which represents the interaction of the measured system with the measurer.

Equation (3.16*a*) for a state and (3.16*b*) for the corresponding projector represent what happens in a single measurement. The density operator equation (3.17) represents what happens in the ensemble of all measurements. The contrast between the nonlinear evolution of individual states represented by the state vector $|\psi\rangle$ or the projector $\mathbf{P}_\psi$ and the linear evolution of the ensemble represented by the density operator $\boldsymbol{\rho}$ was emphasized by Gisin [68], and is treated in more detail in chapter 7.

Although the final density operator can be obtained uniquely from the final states and their probabilities, the final states and probabilities *cannot* be obtained uniquely from the final density operator. Although the final density operator $\boldsymbol{\rho}'$ for the system is diagonal in g, its unravelling into pure states is not unique, and need not consist of eigenstates of $\mathbf{G}$. For a $\boldsymbol{\sigma}_z$ measurement on a two-state system, a possible $\boldsymbol{\rho}'$ and a single state and two possible unravellings are shown in figure 3.2. In a measurement, the final states of the ensemble are all eigenstates of **G**. But one of the unravellings of figure 3.2 is not into eigenstates. The correct unravelling is not given by the density operator alone, but needs further conditions. Thus in a measurement, the final density operator does not give full information about individual states of the ensemble.

Quantum state diffusion for pure states and the master equation for density operators have similar properties, but their evolution is continuous and takes a finite time.

Suppose that an initial ensemble of systems is made up of a mixture of pure states $|\psi_j\rangle$ with probabilities $\mathbf{Pr}(j)$. Then the initial density operator is

$$\rho = \sum_j \mathbf{Pr}(j)|\psi_j\rangle\langle\psi_j| = \sum_j \mathbf{Pr}(j)\mathbf{P}_j. \tag{3.18}$$

The probability that a **G** measurement will give the result g is then

$$\mathbf{Pr}(g) = \sum_j \mathbf{Pr}(j)\mathrm{Tr}\,\mathbf{P}_j\mathbf{P}_g = \mathrm{Tr}\,[(\sum_j \mathbf{Pr}(j)\mathbf{P}_j)\mathbf{P}_g] = \mathrm{Tr}\,[\rho\mathbf{P}_g] \tag{3.19}$$

and the density operator after the measurement is

$$\rho' = \sum_g \mathbf{Pr}(g)\mathbf{P}_g = \sum_g \mathrm{Tr}\,[\rho\mathbf{P}_g]\mathbf{P}_g, \tag{3.20}$$

which is diagonal in **G**-representation.

Both $\mathbf{Pr}(g)$ and ρ' depend linearly on the initial density operator ρ, but do not depend on the way that the density operator is unravelled into pure states with their associated probabilities $\mathbf{Pr}(j)$. Ensembles with the same density operator give the same results for all measurements. The measurements may be delayed, by allowing the ensemble to evolve for a time before making a measurement, either with or without interaction with another system. All such evolutions are linear, with the same result: different ensembles with the same density operator cannot be distinguished by a measurement. That is why this representation is so often used instead of the probability distribution $\mathbf{Pr}(j)$. However, if a system A is entangled with another system B, then measurements on the combined system AB *can* distinguish between different ensembles of A with the same density operator.

The *preparation* of a quantum system is a physical process by which a classical variable influences a quantum state. The definition is used for any such physical process, whether it occurs in the laboratory or not. In a sense, it is the inverse of the measurement process. Section 7.5 shows that there are severe constraints on preparation, due to limitations on the number of distinguishable states of physical systems that are usually treated as classical.

3.6 The boundary problem

The systems of celestial dynamics are usually treated as if they were isolated. When Newton wanted to understand the dynamics of the motion of the Earth (or Earth and Moon) around the Sun, he first ignored the other planets, but recognized that this was an approximation, which was improved much later by including them. The 'system' then took in more than just the Earth and the Sun. The boundary between system and environment was moved outwards to include more of the environment, to improve the approximation.

The same holds for open quantum systems. The system of interest may be the state of excitation of an atom. When the atom is in a cavity, the atom can be treated as the system and the radiation field of the cavity as the environment, leading to exponential radiative decay of excited states. But when the atom is in a highly excited Rydberg state and the cavity is of very high quality, some atomic states may resonate with a mode of the field in the cavity, and the original model is no longer good enough. Modes of the radiation field in the cavity have to be included as part of the system, and the walls of the cavity become the most important part of the environment.

The choice of boundary between system and environment is the *boundary problem*. The atom in a cavity is a practical example. For open quantum systems, entanglement between system and environment makes the boundary problem more subtle and more difficult than for isolated quantum systems or open classical systems.

This is particularly true for the application of quantum state diffusion to measurement, for which the measurer is part of the environment of the system. Where should we put the boundary between the system and the measurer? In the quantum state diffusion theory of measurement this affects our picture of the measurement process. Suppose it is a laboratory measurement of the state of an atom, using emitted photons which are detected using a photomultiplier and counter. The system could be the atom, or it could include part of the photomultiplier, starting with the surface which emits a photoelectron and the electrons which are emitted. When it is just the atom, the measurement is seen as a diffusion process of the state of the atom on the time scale of the decay. But as soon as a part of the photomultiplier is included, the diffusion begins to look like a sudden jump on a far shorter time-scale, as illustrated by various models in the following chapters.

In the practical problems of chapter 6, the ambiguity in the boundary between system and environment can be turned to advantage by a careful choice which includes just enough of the environment to provide the precision required and no more. The boundary problem reappears in the foundations of quantum mechanics as discussed in detail in chapter 7.

3.7 Quantum expectation and quantum variance

The mean value of a dynamical variable g, in the ensemble formed by measuring g in an ensemble of identical pure states $|\psi\rangle$, is the quantum expectation or *expectation*

$$\langle \mathbf{G} \rangle = \langle \mathbf{G} \rangle_\psi = \langle \psi | \mathbf{G} | \psi \rangle, \tag{3.21}$$

where $\mathbf{G}$ is the corresponding operator. The expectation is defined in terms of the operator and initial state before measurement, and is therefore a property of the original pure state. In QSD, both actual means over ensembles, denoted by M, and quantum expectations $\langle . \rangle$ for pure states, which are potential means over ensembles, are important.

Suppose that the current pure state of the system is $|\psi\rangle$ and that $\mathbf{G}$ is an operator. The quantum expectation of $\mathbf{G}$ for the state $|\psi\rangle$ is defined by (3.21). This definition will be used even when $\mathbf{G}$ is not Hermitian, as in the case of creation and annihilation operators.

The corresponding *shifted operator*, which depends on the current state $|\psi\rangle$ and has zero expectation for the current state, is used a lot in QSD theory. It is

$$\mathbf{G}_\Delta = \mathbf{G} - \langle \psi | \mathbf{G} | \psi \rangle = \mathbf{G} - \langle \mathbf{G} \rangle. \tag{3.22}$$

Commutators of shifted and unshifted operators are the same:

$$[\mathbf{G}, \mathbf{F}] = [\mathbf{G}_\Delta, \mathbf{F}]. \tag{3.23}$$

A general operator $\mathbf{G}$ may be separated into its real and imaginary (strictly Hermitian and anti-Hermitian) parts $\mathbf{G}_\mathrm{R}$ and $i\mathbf{G}_\mathrm{I}$, given by

$$\mathbf{G}_\mathrm{R} = \tfrac{1}{2}(\mathbf{G} + \mathbf{G}^\dagger) \qquad i\mathbf{G}_\mathrm{I} = \tfrac{1}{2}(\mathbf{G} - \mathbf{G}^\dagger). \tag{3.24}$$

For the usual annihilation operator $\mathbf{a}$ of an oscillator, these parts are given by

$$\mathbf{x}/(2\hbar)^{\frac{1}{2}}, \qquad i\mathbf{y}/(2\hbar)^{\frac{1}{2}}, \tag{3.25}$$

where $\mathbf{x}$ and $\mathbf{y}$ are conjugate dynamical variables and $\mathbf{a}$, $\mathbf{a}^{\dagger}$, $\mathbf{x}$ and $\mathbf{y}$ satisfy the standard commutation relations,

$$[\mathbf{a}, \mathbf{a}^{\dagger}] = 1, \qquad [\mathbf{x}, \mathbf{y}] = i\hbar. \tag{3.26}$$

Other useful properties of operators and pure states are the quantum variance, which measures the potential variability of an operator $\mathbf{G}$ for a pure state $|\psi\rangle$, and the quantum covariance, which is a bilinear measure of the entanglement, or potential ensemble covariance of two operators $\mathbf{F}$ and $\mathbf{G}$. This is the quantum analogue of the classical covariance, which is a bilinear measure of correlation, and has the same advantages and disadvantages. The quantum variance and covariance are defined in terms of expectations by analogy with the definitions of classical variance and covariance. They can be defined for both Hermitian and non-Hermitian operators.

For an Hermitian operator $\mathbf{G}$ the *quantum variance* $\sigma^2(\mathbf{G})$ for pure state $|\psi\rangle$ is the classical variance of the variable g in the ensemble obtained by measuring $\mathbf{G}$ and $\mathbf{G}^2$. It is

$$\sigma^2(\mathbf{G}) = \langle \mathbf{G}^2 \rangle - \langle \mathbf{G} \rangle^2 = (\Delta g)^2 \qquad \text{(Hermitian)}, \tag{3.27}$$

where Δg is the corresponding standard deviation.

The *quantum covariance* of Hermitian $\mathbf{F}$ and $\mathbf{G}$ is

$$\sigma(\mathbf{F}, \mathbf{G}) = \langle \mathbf{F}\mathbf{G} \rangle - \langle \mathbf{F} \rangle \langle \mathbf{G} \rangle \qquad \text{(Hermitian)}. \tag{3.28}$$

If $\mathbf{F}$ and $\mathbf{G}$ do not commute, then in general $\sigma(\mathbf{F}, \mathbf{G}) \neq \sigma(\mathbf{G}, \mathbf{F})$. The quantum covariance of $\mathbf{G}$ with itself is just the quantum variance of $\mathbf{G}$. In chapter 5 the definitions of quantum variance and covariance are extended to non-Hermitian operators.

Quantum expectation, variance and covariance are potential statistical properties of a pure quantum state, not actual statistical properties. This is an important distinction, particularly when interpreting experiments where systems are entangled. In QSD, ensemble means and variances for ensembles of quantum systems have to be distinguished from the quantum expectations and quantum variances of the individual states.

An open quantum system continually interacts with its environment. Its 'mixed state' is traditionally represented by a density operator $\boldsymbol{\rho}$. In QSD, an individual quantum system is always considered as a pure state represented by a state vector $|\psi\rangle$, whether it is isolated or open, whether it is interacting with its environment or not. The density operator is used, but plays a secondary role.

4

Quantum state diffusion

Chapter 2 showed how Brownian motion, like the dynamics of all open classical systems, can be represented either as the solution of a stochastic equation for an individual system of the ensemble, or in terms of a probability distribution that satisfies a linear deterministic equation. Chapter 3 introduced open quantum systems from the viewpoint of probability distributions and the density operator. This chapter presents the dynamics of open quantum systems, starting in the first section with the linear deterministic master equation for the density operator, and then using it to derive the stochastic quantum state diffusion equations for an individual system of the ensemble. These QSD equations for state vectors are (4.19), (4.20) and (4.21). Measurement and dissipation in two-state systems illustrate the theory by example. Sections 4.4–4.7 present QSD equations for projectors and also alternative stochastic equations that do not satisfy the conditions of QSD. Much of this chapter and the next follows the papers of Gisin and Percival [72, 73, 74, 111].

4.1 Master equations

Interaction of an open quantum system with the environment produces stochastic changes that can be represented by an ensemble of changing pure states $|\psi(t)\rangle$ with probability distribution $\mathbf{Pr}(\psi(t))$, or by a changing density operator $\boldsymbol{\rho}(t)$.

The density operator satisfies a linear master equation. Lindblad [91] showed that if any physical density operator $\boldsymbol{\rho}(t)$ satisfies a linear differential equation, then this equation can always be put into the Lindblad form

$$\dot{\rho} = -\frac{i}{\hbar}[\mathbf{H}, \rho] + \sum_j \left(\mathbf{L}_j \rho \mathbf{L}_j^\dagger - \tfrac{1}{2}\mathbf{L}_j^\dagger \mathbf{L}_j \rho - \tfrac{1}{2}\rho \mathbf{L}_j^\dagger \mathbf{L}_j \right) \qquad \text{(Lindblad)}, \quad (4.1)$$

where the Lindblad operators or *Lindblads* $\mathbf{L}_j$ may or may not be Hermitian.

Master equations have been obtained for a wide variety of open systems, thereby defining Lindblads for these systems. Only the Hamiltonian and Lindblads are needed to formulate QSD for a particular system, and so the experience gained in formulating master equations for open systems can be applied directly to QSD. Measurement is a special case, which is treated in detail later.

Just as the Hamiltonian operator $\mathbf{H}$ determines the contribution of the internal deterministic dynamics to the change of ρ, so the Lindblad operators $\mathbf{L}_j$ determine the contribution of the stochastic dynamics due to interaction with the environment. Later we will see, for example, that an Hermitian Lindblad operator can be used to represent the effect of a measurement, and that a non-Hermitian annihilation Lindblad operator represents a dissipation.

In the isolation limit, when the effect of the environment is negligible, the Lindblads may be neglected, the Hamiltonian remains, and each system of the ensemble evolves by Schrödinger's equation. In the opposite limit, when the effect of the Hamiltonian is negligible and the effect of the environment dominates, the systems are *wide open*, and their evolution is determined by the Lindblad terms. The theory of wide open systems is relatively simple and has many applications. The same theory can often be applied to systems in which the Hamiltonian is important, by using a form of interaction representation to remove the Hamiltonian from the equations.

The simplest type of wide open system is the *elementary open system* with only one Lindblad and with master equation

$$\dot{\rho}(t) = \mathbf{L}\rho(t)\mathbf{L}^\dagger - \tfrac{1}{2}\mathbf{L}^\dagger\mathbf{L}\rho(t) - \tfrac{1}{2}\rho(t)\mathbf{L}^\dagger\mathbf{L}. \qquad (4.2)$$

There are many types of interaction with the environment. We illustrate the properties of master equations and QSD equations using elementary open systems with simple Lindblads representing measurement and dissipation.

In this chapter we take account of the fact that measurement takes a finite time. The measurement of a dynamical variable with Hermitian operator $\mathbf{G}$ is represented by the Lindblad $\mathbf{L} = c\mathbf{G}$, where c is a constant that determines the rate of measurement. Using $|g\rangle$-representation, the solution of the master equation for $\rho_{g'g}(t)$ with initial values $\rho_{g'g}(0)$ is

$$\rho_{g'g}(t) = \exp[-c^2(g' - g)^2 t]\rho_{g'g}(0). \tag{4.3}$$

The off-diagonal elements are suppressed exponentially in time at a rate which increases with the difference between the eigenvalues. When c is large, the measurement takes place rapidly, and tends to become a sudden jump as described in the previous chapter. For a two-state system, the Bloch vector $\vec{r}(t)$ for a $\boldsymbol{\sigma}_z$ measurement with an initial state given by $\vec{r}(0)$ is

$$\vec{r}(t) = (\exp(-c^2 t)x(0),\ \exp(-c^2 t)y(0),\ z(0)). \tag{4.4}$$

The measurement produces an exponential drift along a straight line directly towards the axis joining $|+\rangle$ and $|-\rangle$ states at the north and south poles of the Bloch sphere.

A typical dissipative process is the radiative decay of the excited state of a two-state atom. The excited state is $|e\rangle = |+\rangle$, the ground state is $|g\rangle = |-\rangle$. The decay process is represented by the operator

$$\gamma^{\frac{1}{2}}\boldsymbol{\sigma}_- = \tfrac{1}{2}\gamma^{\frac{1}{2}}(\boldsymbol{\sigma}_x - i\boldsymbol{\sigma}_y) = \gamma^{\frac{1}{2}}\begin{bmatrix} 0 & 0 \\ 1 & 0 \end{bmatrix}, \tag{4.5}$$

whose Hermitian conjugate is $\gamma^{\frac{1}{2}}\boldsymbol{\sigma}_+$. γ is the rate of decay. Also $\boldsymbol{\sigma}_+\boldsymbol{\sigma}_- = \mathbf{P}_+$, the projector onto the excited state. The elementary Lindblad master equation is

$$\dot{\rho} = \gamma\boldsymbol{\sigma}_-\rho\,\boldsymbol{\sigma}_+ - \tfrac{1}{2}\gamma(\mathbf{P}_+\rho + \rho\mathbf{P}_+). \tag{4.6}$$

The general initial condition is a linear combination of the ground and excited states, with Bloch vector $\vec{r}(0)$. The time dependence of the density matrix is

$$\begin{bmatrix} \rho_{++}(0)\,e^{-\gamma t} & \rho_{+-}(0)\,e^{-\gamma t/2} \\ \rho_{-+}(0)\,e^{-\gamma t/2} & \rho_{--}(0)\,(1 - e^{-\gamma t}) \end{bmatrix}. \tag{4.7}$$

The Bloch vector is given by

$$\vec{r}(t) = \left(x(0)\,e^{-\gamma t/2},\ y(0)\,e^{-\gamma t/2},\ (z(0)+1)\,e^{-\gamma t} - 1\right) \tag{4.8}$$

and moves along a parabolic path to the south pole, which represents the ground state.

The master equation for an elementary system is invariant under multiplication of $\mathbf{L}$ by a complex phase factor u of unit modulus. The same effect is produced by $\mathbf{L}$ and by $u\mathbf{L}$, even when $\mathbf{L}$ is Hermitian and $u\mathbf{L}$ is not. For elementary systems, one Lindblad equation uniquely represents the evolution of the ensemble.

The Lindblad equation for a general system is similarly invariant for multiplication of the vector $\vec{\mathbf{L}}$ of Lindblads $\mathbf{L}_j$ by a unitary matrix u_{kj}. The new equation, obtained by replacing the $\mathbf{L}_j$ by $\mathbf{L}'_k = \sum u_{kj}\mathbf{L}_j$ is also of Lindblad form. It may be considered as a different but equivalent equation, but it is preferable to consider it as the same equation with a different representation for the Lindblad vector in a linear space. The QSD equations also have this unitary invariance property, which is not shared by all stochastic diffusion equations for quantum states.

The ambiguity in the choice of Lindblads is a nuisance, so it is helpful to adopt a convention which reduces the choice. The convention is that where possible Hermitian Lindblads are used in preference to others, giving a standard form for this case.

4.2 QSD equations from master equations

An ensemble of quantum systems, whose state vectors satisfy a stochastic differential equation, has a density operator that satisfies a unique deterministic differential master equation. By contrast a master equation corresponds to many different stochastic equations for its component states. These are known as different *unravellings* of the master equation. QSD is one such unravelling.

This section shows how the quantum state diffusion equation is derived from the master equation (4.1) for an elementary system.

A density operator ρ for an evolving quantum system can be expressed in many ways as a mean M over a distribution of normalized pure state projection operators. We seek differential equations for $|\psi(t)\rangle$ such that the density operator given by the ensemble mean over the projectors

$$\rho(t) = \mathrm{M}\,|\psi(t)\rangle\langle\psi(t)| \tag{4.9}$$

satisfies the Lindblad master equation (4.1).

In time $\mathrm{d}t$ the variation $|\mathrm{d}\psi\rangle$ in $|\psi\rangle$ is then given by the Itô equation

$$|\mathrm{d}\psi\rangle = |\mathrm{drift}\rangle\mathrm{d}t + |\mathrm{fluct}\rangle\mathrm{d}\xi = |v\rangle\mathrm{d}t + |f\rangle\mathrm{d}\xi. \tag{4.10}$$

The evolution of the state vector is continuous. The differential stochastic fluctuation $\mathrm{d}\xi$ has equal and independent fluctuations in its real and imaginary parts (chapter 2):

$$\mathrm{M}\,\mathrm{d}\xi = 0, \qquad \mathrm{M}\,(\mathrm{d}\xi)^2 = 0, \qquad \mathrm{M}\,|\mathrm{d}\xi|^2 = \mathrm{d}t. \tag{4.11}$$

Fluctuations at different times are assumed to be statistically independent, so the stochastic differential equation represents a Markov process. The state vector remains normalized. These four conditions lead to a unique unravelling into states whose trajectories in Hilbert space satisfy the QSD equations. The other unravellings described in sections 4.5 and 4.6 fail to satisfy at least one of these conditions.

First we derive the dominant fluctuation terms, and then the drift. To preserve normalization, keeping the state vector on the unit sphere, the differential change in the state vector due to the fluctuation must be orthogonal to the state vector itself, because the fluctuations are of higher order than the drift. So

$$\langle\psi|f\rangle = 0. \tag{4.12}$$

Taking means over $|\mathrm{d}\psi\rangle$ and $|\mathrm{d}\psi\rangle\langle\mathrm{d}\psi|$ in equation (4.10),

$$\left.\begin{aligned} \mathrm{M}\,|\mathrm{d}\psi\rangle &= |v\rangle \mathrm{d}t \\ \mathrm{M}\,|\mathrm{d}\psi\rangle\langle\mathrm{d}\psi| &= |f\rangle\langle f|\mathrm{d}t, \end{aligned}\right\} \tag{4.13}$$

so the change in the density operator is given by

$$\begin{aligned} \mathrm{d}\rho &= \mathrm{M}\,(|\psi\rangle\langle\mathrm{d}\psi| + |\mathrm{d}\psi\rangle\langle\psi| + |\mathrm{d}\psi\rangle\langle\mathrm{d}\psi|) \\ \text{or} \qquad \dot{\rho} &= (|\psi\rangle\langle v| + |v\rangle\langle\psi| + |f\rangle\langle f|). \end{aligned} \tag{4.14}$$

The stochastic terms are determined by the component of $\dot{\rho}$ in the space orthogonal to $|\psi\rangle$, so, using the master equation (4.1),

$$\begin{aligned} |f\rangle\langle f| &= (\boldsymbol{I} - |\psi\rangle\langle\psi|)\,\dot{\rho}\,(\boldsymbol{I} - |\psi\rangle\langle\psi|) = (\boldsymbol{I} - |\psi\rangle\langle\psi|)\,\mathbf{L}\rho\mathbf{L}^\dagger\,(\boldsymbol{I} - |\psi\rangle\langle\psi|) \\ &= (\boldsymbol{I} - |\psi\rangle\langle\psi|)\,\mathbf{L}|\psi\rangle\langle\psi|\,\mathbf{L}^\dagger(\boldsymbol{I} - |\psi\rangle\langle\psi|) \\ &= |(\mathbf{L} - \langle\mathbf{L}\rangle)\psi\rangle\langle(\mathbf{L} - \langle\mathbf{L}\rangle)\psi| = |\mathbf{L}_\Delta\psi\rangle\langle\mathbf{L}_\Delta\psi|, \end{aligned} \tag{4.15}$$

where $\langle\mathbf{L}\rangle$ and $\mathbf{L}_\Delta$ are defined by (3.21) and (3.22). So

$$|f\rangle = u(\mathbf{L} - \langle\mathbf{L}\rangle)|\psi\rangle, \tag{4.16}$$

where u is a phase factor for the operator $\mathbf{L}$. However in the QSD equation (4.10) the distribution of the fluctuations $d\xi$ given by (4.11) is invariant under multiplication by a phase factor, so that u makes no difference to the stochastic equations and can be omitted.

The drift is given by

$$\dot{\rho}|\psi\rangle = |\psi\rangle\langle v|\psi\rangle + |v\rangle, \tag{4.17}$$

from which it follows on premultiplying by $\langle\psi|$ that

$$\begin{aligned} \mathrm{Re}\langle\psi|v\rangle &= \tfrac{1}{2}\langle\psi|\dot{\rho}|\psi\rangle \\ \text{so that} \qquad |v\rangle &= \dot{\rho}|\psi\rangle - \tfrac{1}{2}\langle\psi|\dot{\rho}|\psi\rangle|\psi\rangle + i\eta(t)|\psi\rangle \\ &= \left(\langle\mathbf{L}^\dagger\rangle\mathbf{L} - \tfrac{1}{2}\mathbf{L}^\dagger\mathbf{L} - \tfrac{1}{2}\langle\mathbf{L}^\dagger\rangle\langle\mathbf{L}\rangle + i\eta(t)\right)|\psi\rangle, \end{aligned} \tag{4.18}$$

where the last line follows from the master equation (4.1) with $\rho = |\psi\rangle\langle\psi|$. The pure imaginary function of time $i\eta(t)$ changes only the phase factor of the state vector, and so is nonphysical. It is chosen by convention to be zero, so that the phase change agrees with the Schrödinger equation in the absence of interaction with the environment.

So the elementary QSD equation is

$$\left.\begin{aligned} |d\psi\rangle &= \left(\langle\mathbf{L}^\dagger\rangle\mathbf{L} - \tfrac{1}{2}\mathbf{L}^\dagger\mathbf{L} - \tfrac{1}{2}\langle\mathbf{L}^\dagger\rangle\langle\mathbf{L}\rangle\right)|\psi\rangle dt + (\mathbf{L} - \langle\mathbf{L}\rangle)|\psi\rangle d\xi \\ &= \left(\langle\mathbf{L}^\dagger\rangle\mathbf{L} - \tfrac{1}{2}\mathbf{L}^\dagger\mathbf{L} - \tfrac{1}{2}\langle\mathbf{L}^\dagger\rangle\langle\mathbf{L}\rangle\right)|\psi\rangle dt + \mathbf{L}_\Delta|\psi\rangle d\xi. \end{aligned}\right\} \tag{4.19}$$

For Hermitian $\mathbf{L}$ this becomes

$$|d\psi\rangle = -\tfrac{1}{2}\mathbf{L}_\Delta^2|\psi\rangle dt + \mathbf{L}_\Delta|\psi\rangle d\xi, \tag{4.20}$$

where $\mathbf{L}_\Delta = \mathbf{L} - \langle\mathbf{L}\rangle$ is the shifted $\mathbf{L}$-operator whose expectation for the current state $|\psi\rangle$ is zero. This QSD equation represents the measurement of a dynamical variable whose operator is proportional to $\mathbf{L}$. It can be shown that the mean of the expectation $\mathrm{M}\,\langle\mathbf{L}\rangle$ remains constant, as would be expected from a measurement.

For general systems, the derivation of the QSD equation follows the same lines, with a unitary transformation of the $\mathbf{L}_j$ instead of the phase factor u in

(4.16). Given the Lindblad master equation (4.1) for the density operator $\boldsymbol{\rho}$, the stochastic differential equation for the state vector $|\psi\rangle$ is

$$\begin{aligned}|d\psi\rangle = & -\frac{i}{\hbar}\mathbf{H}|\psi\rangle dt + \sum_j \Big(\langle \mathbf{L}_j^\dagger\rangle \mathbf{L}_j - \tfrac{1}{2}\mathbf{L}_j^\dagger \mathbf{L}_j - \tfrac{1}{2}\langle \mathbf{L}_j^\dagger\rangle\langle \mathbf{L}_j\rangle\Big)|\psi\rangle dt \\ & + \sum_j (\mathbf{L}_j - \langle \mathbf{L}_j\rangle)|\psi\rangle d\xi_j.\end{aligned} \tag{4.21}$$

The general equation follows from Schrödinger's equation and the equation (4.19) for an elementary system on using the additivity rule of section 2.9. Because of the rule, this book concentrates throughout on elementary systems with no Hamiltonian and only one Lindblad.

The first sum in (4.21) represents the drift of the state vector and the second sum the independent random fluctuations due to the different interactions of the system with its environment. $\langle \mathbf{L}_j\rangle = \langle\psi|\mathbf{L}_j|\psi\rangle$ is the expectation of $\mathbf{L}_j$ for state $|\psi\rangle$, and the density operator is given by the mean over the projectors onto the quantum states of the ensemble:

$$\boldsymbol{\rho} = \mathrm{M}\,|\psi\rangle\langle\psi|, \tag{4.22}$$

where M represents a mean over the ensemble.

The $d\xi_j$ are independent normalized complex differential random variables, satisfying the complex relations from (2.33), which are

$$\mathrm{M}\,d\xi_j = 0, \tag{4.23a}$$

$$\mathrm{M}\,(d\xi_j d\xi_{j'}) = 0, \qquad \mathrm{M}\,(d\xi_j^* d\xi_{j'}) = \delta_{jj'}dt. \tag{4.23b}$$

Unlike the master equation, the QSD equation (4.21) is *nonlinear*, because the scalars $\langle \mathbf{L}_j\rangle$ depend quadratically on $|\psi\rangle$. If the time dependence of the scalars is known, then the state can be obtained by solving a linear stochastic differential equation, suggesting a possible iterative solution.

The QSD equations are also usually *nonlocal*, which means that the solutions depend on distant influences. This can be seen for the case of the elementary open system whose state is represented by a wave function $\psi(x, t)$. Suppose the Lindblad in the QSD equation (4.19) is the position operator $\mathbf{L} = \mathbf{x}$. The value of $\psi(x', t + dt)$ for a particular point x' depends on the expectation $\langle \mathbf{x}\rangle = \int dx \cdot |\psi(x)|^2 x$, which depends on all the non-zero

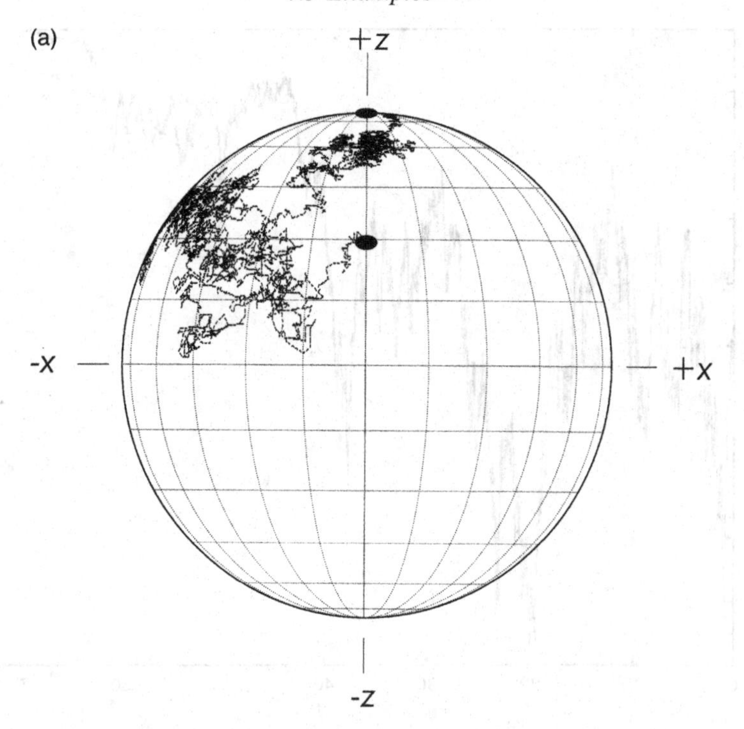

Figure 4.1a A single trajectory on the Bloch sphere for $\mathbf{H} = 0,\ \mathbf{L} = c\sigma_z$.

values of $\psi(x, t)$, however far from x' the point x may be. This nonlocality causes problems for relativistic QSD, as discussed in section 8.7.

The solution of a QSD equation is a nondifferentiable function of time. Its Fourier transform is like the Fourier transform of white noise, whose amplitudes are significant up to arbitrarily high frequencies. Bohr's relation $E = \hbar\omega$ implies correspondingly high energies, but there are no such energies, so the usual Fourier relation between energy and time breaks down in nonrelativistic QSD. But the Fourier relation between momentum and position is retained, introducing a fundamental asymmetry that is incompatible with special relativity.

4.3 Examples

Consider the examples of two-state measurement and dissipation. In each case the QSD projector diffuses on the surface of the Bloch sphere. For a $\boldsymbol{\sigma}_z$ measurement, the diffusion is illustrated in figure 4.1a. Figure 4.1b illustrates the time dependence of $\langle\boldsymbol{\sigma}_z\rangle$ for the same trajectory.

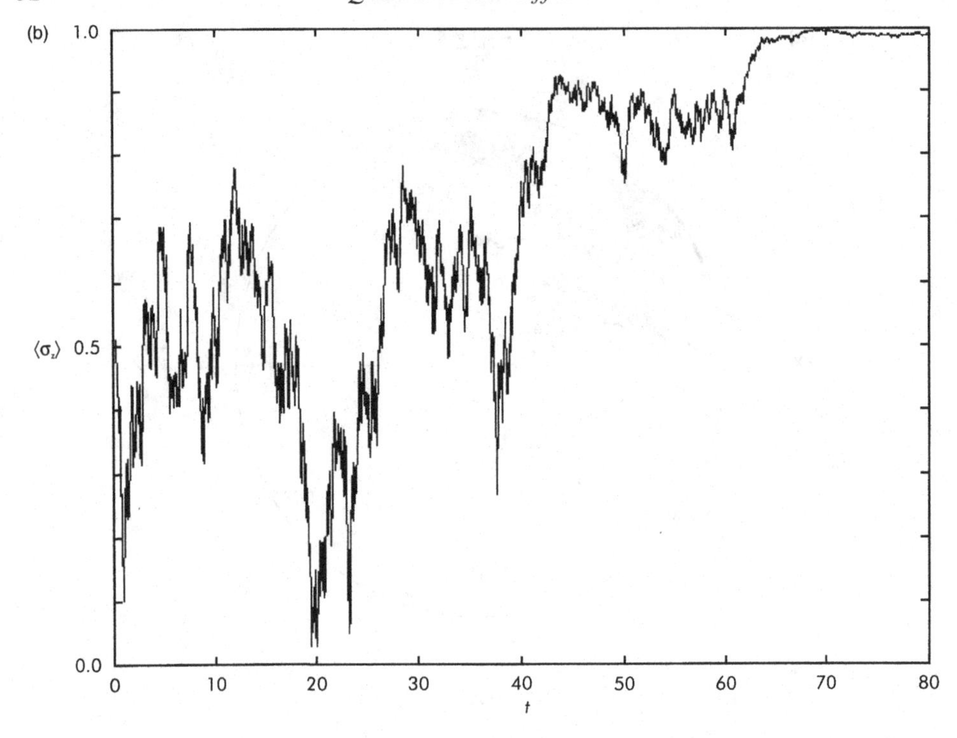

Figure 4.1b The time dependence of $\langle \sigma_z \rangle$.

The projector diffuses around the surface of the sphere to either the north or south pole, corresponding to the $|+\rangle$ or $|-\rangle$ state of the system. The probability of diffusion to each of these states from an initial state $|\psi(0)\rangle$ is given by initial expectations of the corresponding projectors. This is an illustration of the expectations for states becoming probabilities for the ensemble after the measurement. This is shown for a small sample in figure 4.2.

Notice that for QSD there is no ambiguity in the unravelling. The ensemble of trajectories is defined uniquely. The equations not only give the correct evolution of the density operator, which was an input into their derivation, but they also give the unravelling of the density operator into evolving pure states, all of which approach the eigenstates $|+\rangle$ and $|-\rangle$. This long-term behaviour of individual systems was not put in as a condition at any stage. Of course it is not possible to predict which of the trajectories of the ensemble will be seen in a given experiment. That depends on the fluctuation, which is unknown.

It may seem surprising that diffusing states concentrate at two points on the sphere. It is as if the sphere becomes more and more sticky as the state

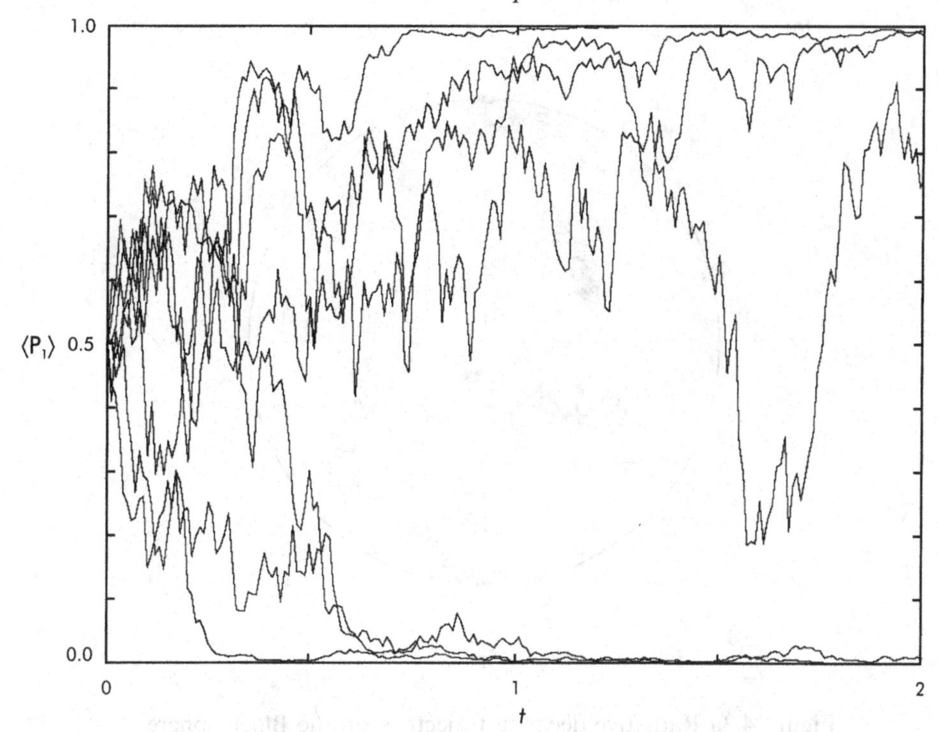

Figure 4.2 The time dependence of $\langle \mathbf{P}_1 \rangle$ for seven trajectories with the same initial condition $|\psi(0)\rangle = (1/\sqrt{2})(|+\rangle + |-\rangle)$, but different fluctuations.

approaches an eigenstate, so that the states of the system find it more difficult to diffuse away than to diffuse towards the eigenstates.

Figures 4.3a and 4.3b illustrate the radiative decay of a two-state atom from the excited state. The diffusion typically takes a time comparable to the decay time, with a significant proportion of this time in intermediate states between the excited and ground states.

This is not seen in the laboratory, where the emitted photons can be counted, and sudden quantum jumps are seen. Such jumps are also seen in QSD as a result of simultaneous measurement and radiative decay, which is the correct QSD representation of the decay process as seen in the laboratory. The QSD equation is no longer elementary, but has two Lindblad operators,

$$\mathbf{L}_1 = \gamma\, \boldsymbol{\sigma}_- \qquad \text{(decay)}, \qquad \mathbf{L}_2 = c\, \boldsymbol{\sigma}_z \qquad \text{(measurement)}. \qquad (4.24)$$

The time dependence of $\langle \mathbf{P}_1 \rangle$ is illustrated in figures 4.4a–c for $\gamma = 0.3$, $c = 0.15, 0.3$ and 8. As the measurement Lindblad increases, the continuous

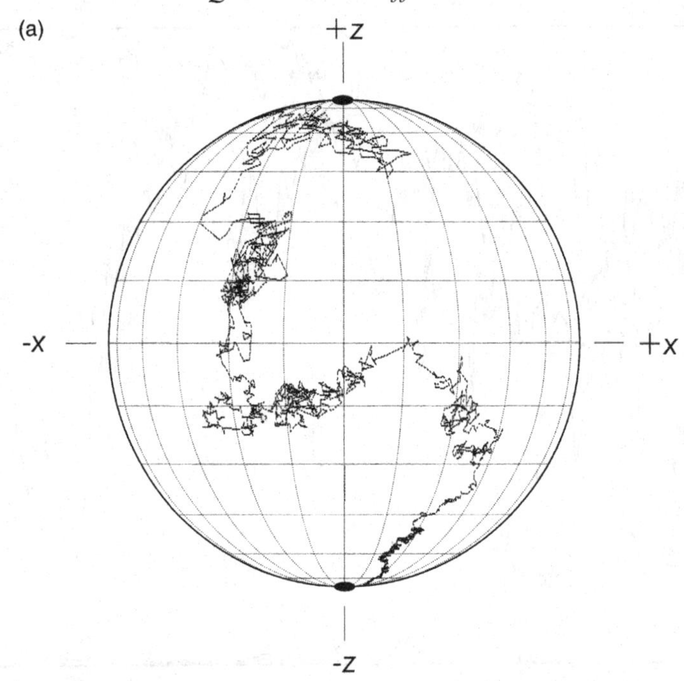

Figure 4.3a Radiative decay: a trajectory on the Bloch sphere.

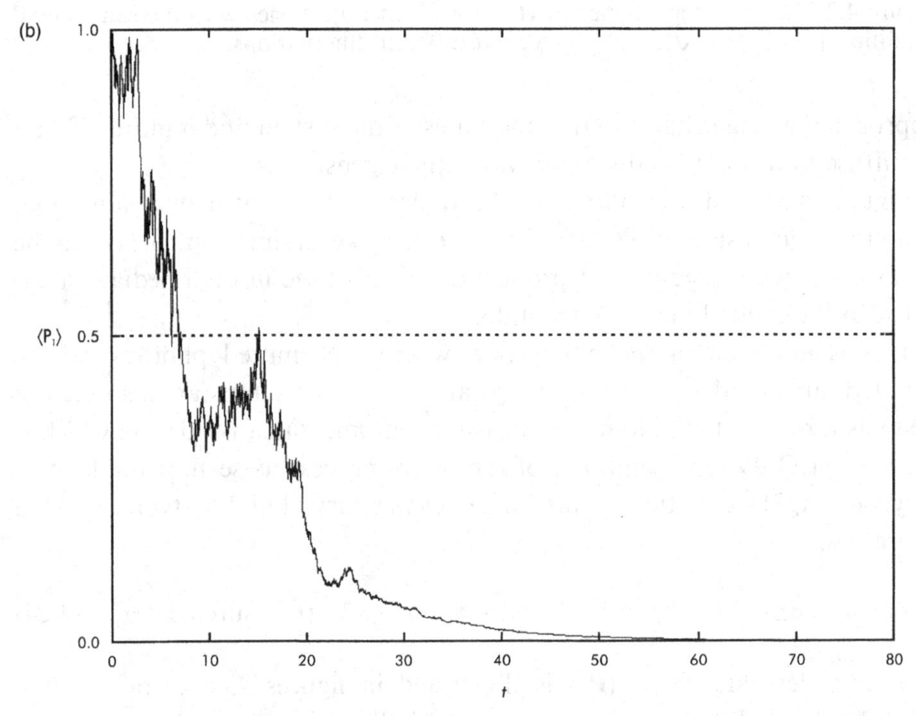

Figure 4.3b The time dependence of the expectation of the projector $\langle \mathbf{P}_1 \rangle = \frac{1}{2}(1 + \langle \boldsymbol{\sigma}_z \rangle)$, which is +1 for the excited state, and 0 for the ground state.

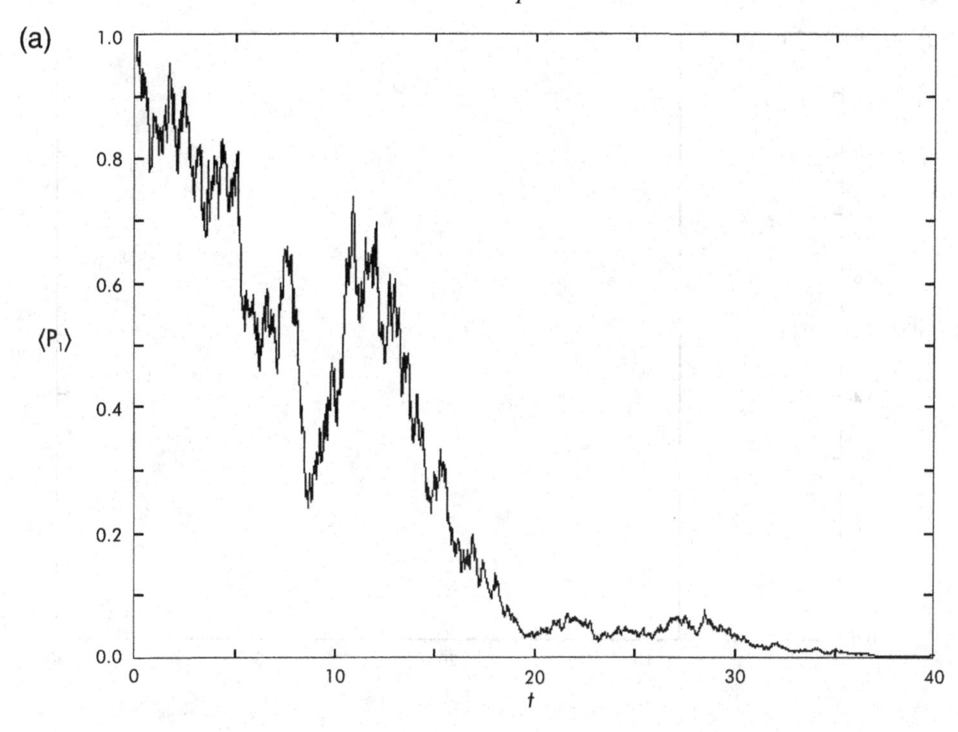

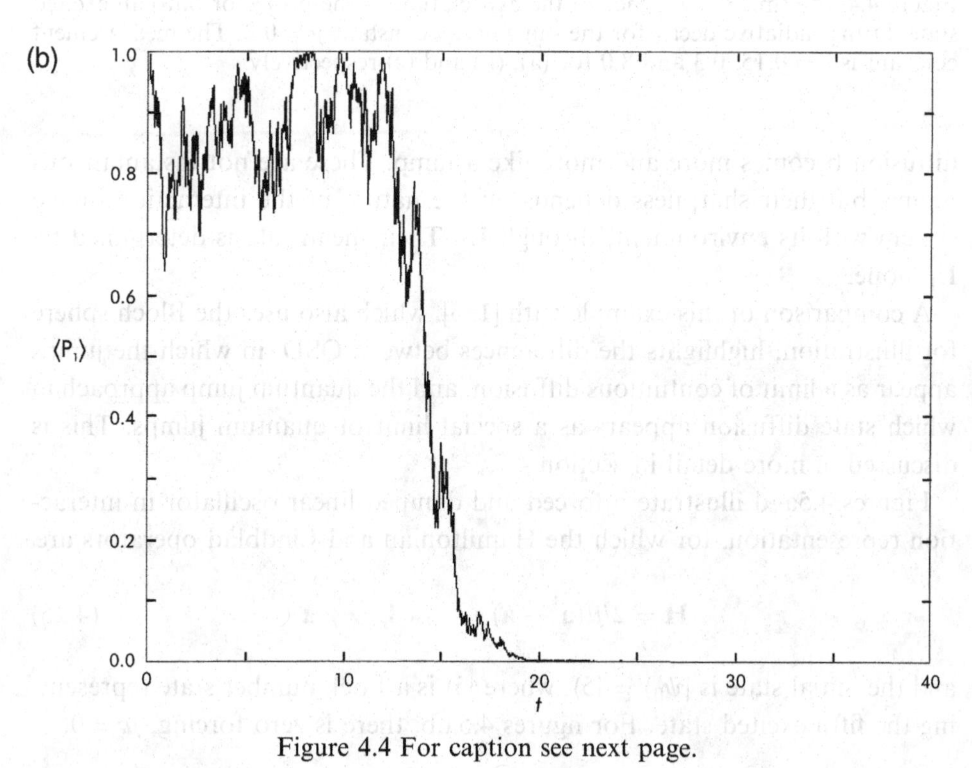

Figure 4.4 For caption see next page.

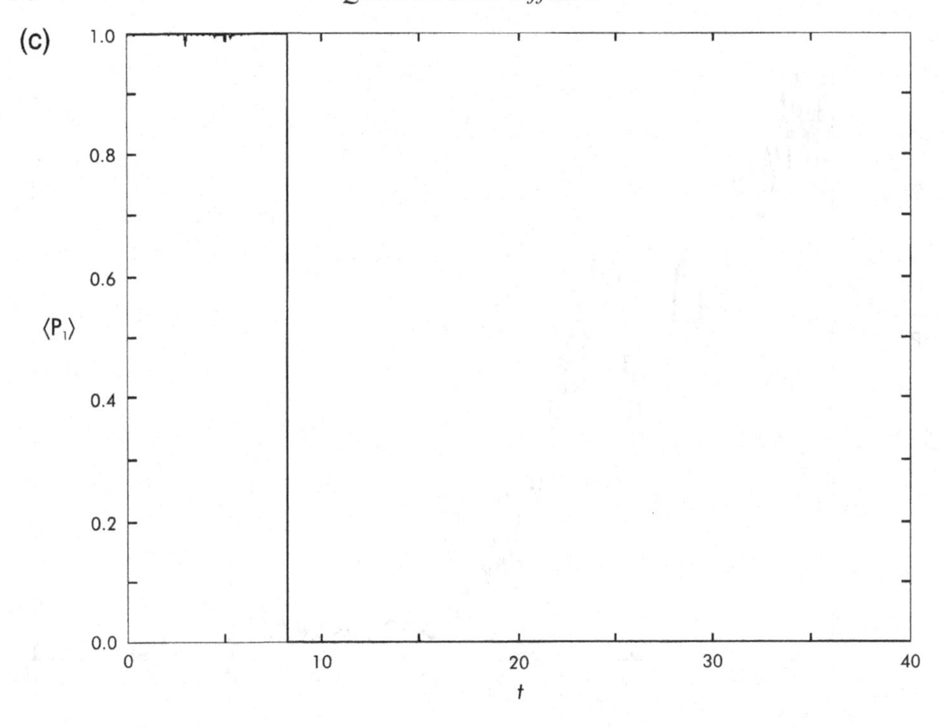

Figure 4.4 The time dependence of the expectation of the projector onto an excited state during radiative decay for the same decay constant $\gamma = 0.3$. The measurement constant is $c = 0.15$, 0.3 and 8.0 for (a), (b) and (c) respectively.

diffusion becomes more and more like a jump. These are not instantaneous jumps, but their sharpness depends on the nature of the interaction of the system with its environment, through $\mathbf{L}_2$. Their mean rate is determined by $\mathbf{L}_1$ alone.

A comparison of this example with [155], which also uses the Bloch sphere for illustration, highlights the differences between QSD, in which the jumps appear as a limit of continuous diffusion, and the quantum jump approach in which state diffusion appears as a special limit of quantum jumps. This is discussed in more detail in section 4.7.

Figures 4.5a–d illustrate a forced and damped linear oscillator in interaction representation, for which the Hamiltonian and Lindblad operators are

$$\mathbf{H} = 2i\mu(\mathbf{a}^\dagger - \mathbf{a}), \qquad \mathbf{L} = \gamma\,\mathbf{a} \tag{4.25}$$

and the initial state is $|\psi_0\rangle = |5\rangle$, where $|5\rangle$ is a Fock number state representing the fifth excited state. For figures 4.5a,b, there is zero forcing, $\mu = 0$.

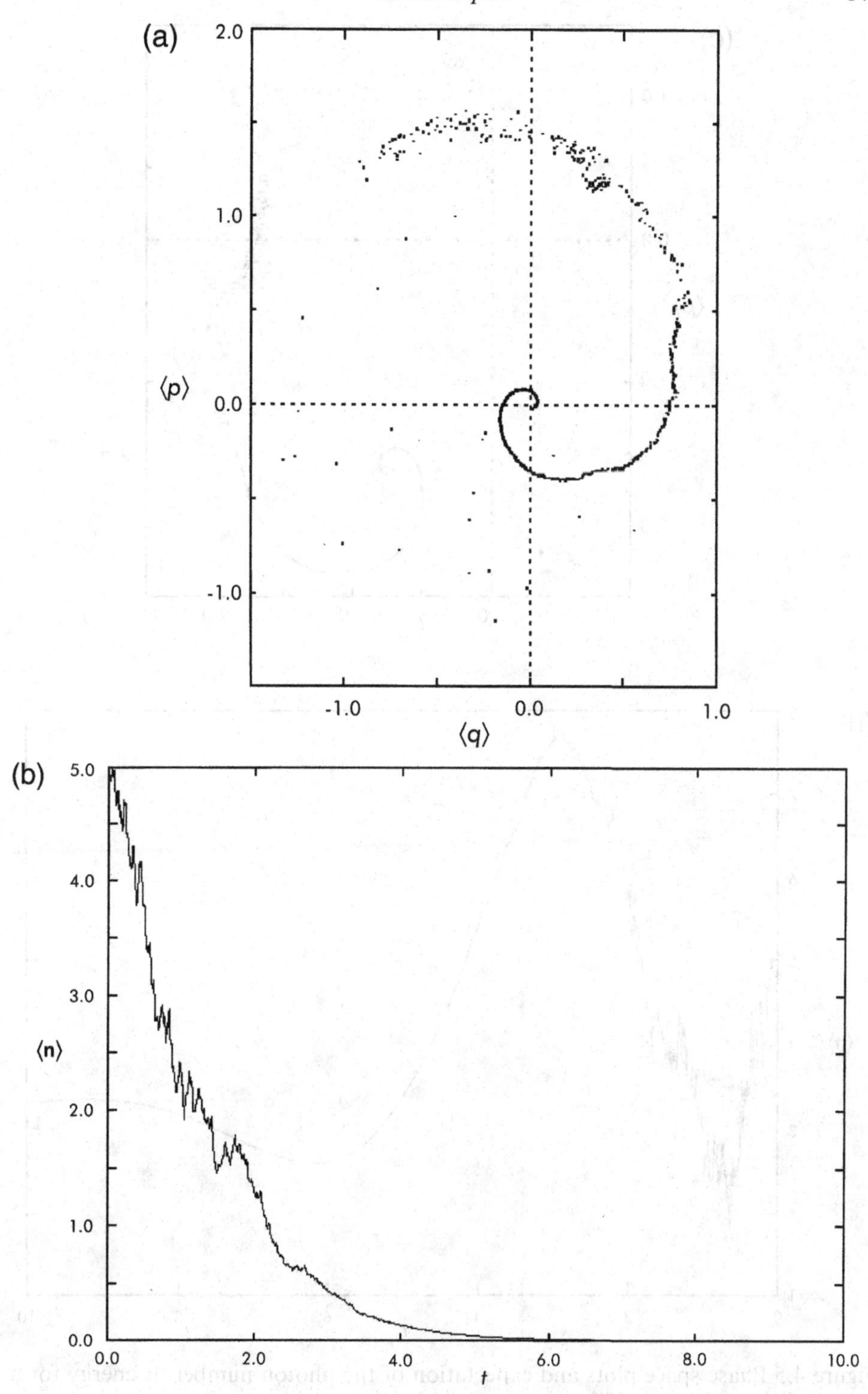

Figure 4.5 For caption see next page.

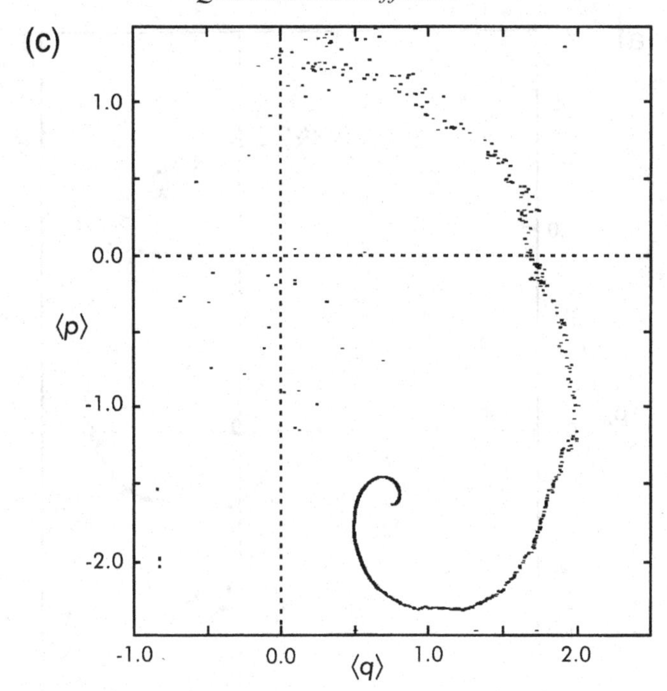

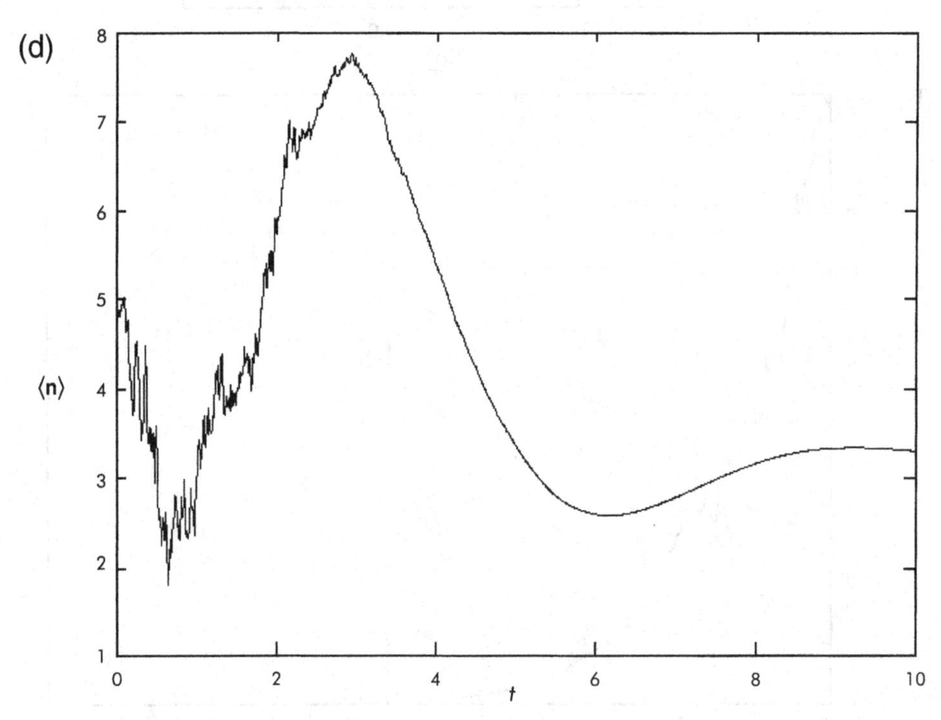

Figure 4.5 Phase space plots and expectation of the photon number or energy for a forced, damped linear oscillator.

The figures illustrate state diffusion towards a coherent state, which moves like a classical particle and has no stochastic fluctuations. This is an example in the quantum domain of localization in phase space, which is characteristic of our everyday experience in the classical domain. We use *localization* in the sense of the state of an individual system of an ensemble being confined to the neighbourhood of a single state or group of states. We use it for a quantum state being localized in the neighbourhood of a point in phase space, as for a confined wave packet. We also use it to describe the processes that lead to these localized states, sometimes called *reduction* in the traditional theory of quantum measurement.

The next example is a quantum cascade for an oscillator with emission and measurement, also in interaction representation. The operator $\mathbf{L}_1$ represents the interaction required to measure the system in a given state, whereas $\mathbf{L}_2$ represents the damping due to photon emission. The operators are

$$\mathbf{H} = 0, \qquad \mathbf{L}_1 = 6\mathbf{a}^\dagger\mathbf{a}, \qquad \mathbf{L}_2 = 0.1\mathbf{a}. \tag{4.26}$$

Figure 4.6 shows that the continuous state diffusion approximates to a succession of sudden jumps.

Our final example, illustrated in figure 4.7, shows how a Hamiltonian can turn momentum localization into phase space localization, with a resultant position localization. This is a very simple example of a general phenomenon for systems of many freedoms, in which the dynamics of the Hamiltonian can produce phase space localization from a few simple Hermitian Lindblads, even just one. The operators are

$$\mathbf{H} = \tfrac{1}{2}(\mathbf{q}^2 + \mathbf{p}^2), \qquad \mathbf{L} = 0.3\mathbf{p}. \tag{4.27}$$

The initial condition is the Fock number state $|5\rangle$ and the graph shows $\langle\mathbf{q}\rangle$ and $\langle\mathbf{p}\rangle$ as a function of time, with quantum standard deviations, the square root of the variance, for each. The Hamiltonian causes a rotation in phase space, which turns a wave packet that is localized in momentum into a wave packet localized in position, with the result that both of them localize, though not quite to a minimum indeterminacy wave packet for these conjugate variables.

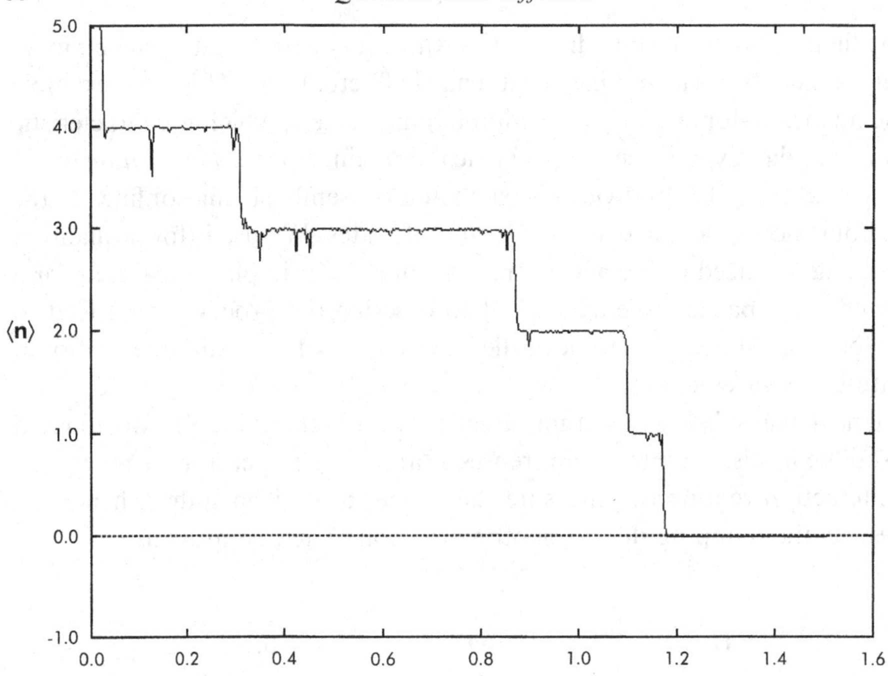

Figure 4.6 Sudden jumps approximated by continuous state diffusion.

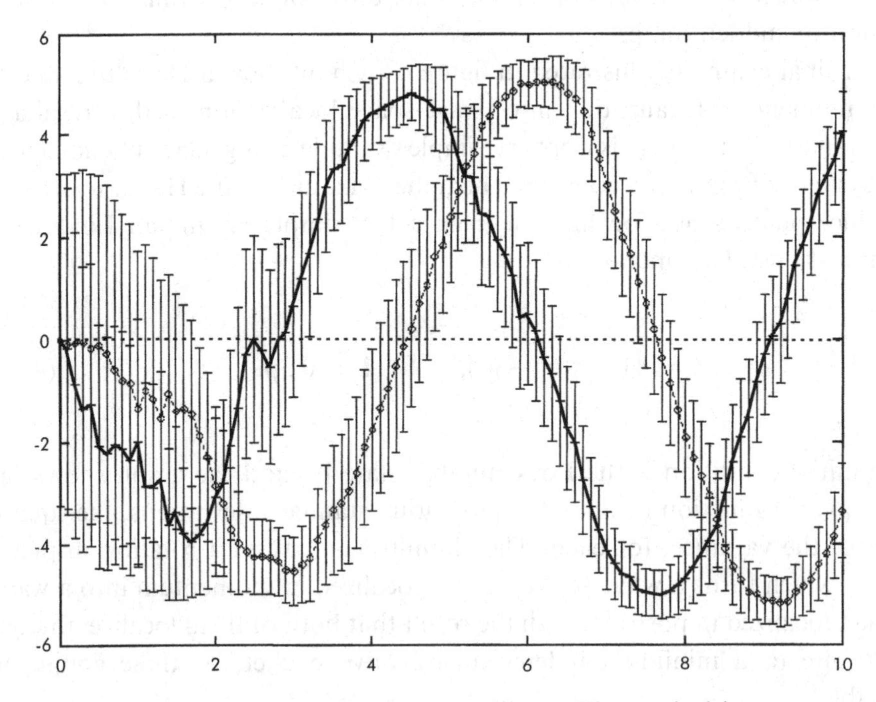

Figure 4.7 Momentum localization for an oscillator. The vertical bars are Δq and Δp. The graph of $\langle \mathbf{p} \rangle$ is distinguished by the continuous line at the centre of the bars.

Other examples are given in the papers listed at the end of the chapter summary, including symmetry breaking by localization and a QSD picture of measurement by the formation of a photographic latent image. Spiller and his collaborators have studied thermal equilibrium and the behaviour of quantum capacitors [139, 137]. Further examples are given in more detail in chapter 6.

4.4 Projectors

Quantum state diffusion has a particularly elegant representation in terms of projection operators, or projectors [34]. It follows directly from the fluctuation term $\mathbf{L}_{\Delta}|\psi\rangle d\xi$ in the elementary QSD equation (4.19) that the equation of change for the pure state projector has the form

$$d\mathbf{P}_{\psi} = d(|\psi\rangle\langle\psi|) = \mathbf{X}dt + \mathbf{L}_{\Delta}\mathbf{P}_{\psi}d\xi + \mathbf{P}_{\psi}\mathbf{L}_{\Delta}^{\dagger}d\xi^{*} \tag{4.28}$$

and it follows from the master equation (4.1) for the ensemble with initial density operator $\mathbf{P}_{\psi}$ that

$$\left.\begin{aligned} \mathbf{X}dt &= \mathbf{M}\,d\mathbf{P}_{\psi} = d\rho \\ &= \mathbf{L}\mathbf{P}_{\psi}\mathbf{L}^{\dagger}dt - \tfrac{1}{2}(\mathbf{L}^{\dagger}\mathbf{L}\mathbf{P}_{\psi} + \mathbf{P}_{\psi}\mathbf{L}^{\dagger}\mathbf{L})dt, \end{aligned}\right\} \tag{4.29}$$

giving

$$\begin{aligned} d\mathbf{P}_{\psi} &= \mathbf{L}\mathbf{P}_{\psi}\mathbf{L}^{\dagger}dt - \tfrac{1}{2}(\mathbf{L}^{\dagger}\mathbf{L}\mathbf{P}_{\psi} + \mathbf{P}_{\psi}\mathbf{L}^{\dagger}\mathbf{L})dt \\ &\quad + \mathbf{L}_{\Delta}\mathbf{P}_{\psi}d\xi + \mathbf{P}_{\psi}\mathbf{L}_{\Delta}^{\dagger}d\xi^{*} \qquad \text{(elementary).} \end{aligned} \tag{4.30}$$

The generalization from the elementary equations for one Lindblad and no Hamiltonian to the general equations with a Hamiltonian and any number of Lindblads is obvious.

In this representation the connection between the QSD equation for the state vector and the master equation for the density operator is particularly clear.

4.5 Linear unravelling

If we just omit the nonlinear terms from the QSD equations we get the much simpler linear equations

$$|d\phi\rangle = -\tfrac{1}{2}\mathbf{L}^\dagger\mathbf{L}|\phi\rangle dt + \mathbf{L}|\phi\rangle d\xi. \tag{4.31}$$

The linear form of the projector equations is exactly the same as above (4.30), with $\mathbf{L}$ in place of $\mathbf{L}_\Delta$, and it is obvious that the mean gives the usual master equations. So why isn't the linear form always used? It is because the norm of the individual state wave functions is not preserved. The change in the norm is given by

$$\left.\begin{aligned} d(|\phi|^2) = \mathrm{Tr}\, d\mathbf{P}_\phi &= \langle\phi|\mathbf{L}|\phi\rangle d\xi + \langle\phi|\mathbf{L}^\dagger|\phi\rangle d\xi^* \\ &= 2\langle\phi|\mathbf{L}_R|\phi\rangle d\xi_R - 2\langle\phi|\mathbf{L}_I|\phi\rangle d\xi_I. \end{aligned}\right\} \tag{4.32}$$

For linear unravelling the ensemble is not a normal ensemble in which each member of the ensemble has equal weight, but a *weighted* ensemble, in which the weight is given by the changing norm. Linear unravelling has many advantages, because the analysis is much simpler. In particular the path integral formulation is relatively straightforward for linear unravelling.

But it also has a major disadvantage. The $|\phi(t)\rangle$ of linear unravelling does not represent an individual physical system. Even if each solution of the stochastic equations is normalized, giving

$$|\phi'(t)\rangle = |\phi(t)\rangle/|\phi(t)|, \tag{4.33}$$

the normalized states are *not* solutions of the ordinary QSD equations, their mean does not give the master equation and they do not represent the evolution of an individual system. Thus a solution of the linear equations cannot be used to represent the evolution of an individual system, such as a single run of a laboratory experiment.

A consequence of this is that when the solutions of the linear stochastic equations are used in a Monte Carlo numerical method, the sample tends to be dominated by the single member with the greatest norm, so that the statistics are very bad when the runs last a significant period of time. The statistics can be improved by simulating biological reproduction of the members with large norm, but this complicates the programming, and is still inferior to the usual nonlinear numerical Monte Carlo method described in chapter 6. Another consequence is that the linear stochastic equations do not satisfy Bell's conditions for a good quantum theory in chapter 7.

Hudson and Parthasarathy have developed the mathematics of linear unravelling [85, 102], whereas Goetsch and Graham treat the physics [79].

4.6 Other fluctuations

We now return to the nonlinear theory. The unitary condition described after equation (4.8) is a consequence of having complex fluctuations satisfying the conditions (4.11) or (4.23). We now relax these conditions, allowing other types of fluctuation. Since these conditions are needed to obtain a unique QSD equation from a master equation, there is then a large choice of possible quantum state diffusion equations consistent with a given master equation.

The two most important can be expressed in terms of real fluctuations dw defined in section 2.9. They can be illustrated for a two-state system with the Hermitian Lindblad $\mathbf{L} = \boldsymbol{\sigma}_z$ on the Bloch sphere of figure 3.1. First suppose that $d\xi$ is replaced by dw in the fluctuation term. Then the state diffuses along a line of longitude to either the north or south pole of the Bloch sphere. Like QSD, this state diffusion process localizes, and represents a measurement of the z-component of the spin. Such state diffusion equations can be used to represent measurement. They have been used extensively by researchers on continuous spontaneous localization, discussed in more detail in chapter 7.

Secondly suppose that $d\xi$ is replaced by $-idw$, so that the fluctuations are pure imaginary. Then, for the two-state system with Lindblad $\mathbf{L} = \boldsymbol{\sigma}_z$, the state diffuses along a line of latitude. There is no localization, and the theory cannot be used to represent a measurement. This form of state diffusion represents a system with a fluctuating Hamiltonian, for example a charged particle or atom in a classical electromagnetic field. Ordinary QSD with complex fluctuations is used to represent *interactions* between system and environment, which produces entanglement between them. The pure imaginary fluctuations represent the *action* of a stochastic environment on the system, represented by a fluctuating Hamiltonian, which does not produce entanglement.

The general form of the state diffusion equations with real and pure imaginary fluctuations is the same as for the QSD equations, with $d\xi$ replaced by dw or by idw. The measurement and the fluctuation result in the same master equation and the same evolution of the density operator. So the two processes cannot be distinguished from each other or from QSD by looking at the density operator. However, they represent different evolutions for the individual systems. For both real and imaginary fluctuations, it must be remembered that the theory is no longer invariant under unitary transformations in the space of the fluctuations, nor in the space of the Lindblad operators. Consequently, phase factors in these operators now make a difference to the evolution of the individual states of the ensemble.

4.7 QSD, jumps and Newtonian dynamics

So far we have assumed that the quantum states satisfy a stochastic differential equation with continuous solutions. Not all unravellings have this property. There is also the possibility that the stochastic process consists of a sequence of discrete quantum jumps. Such unravellings are very commonly used for the solution of practical problems in quantum optics, and they have also been used as the foundation of an alternative quantum theory.

In comparing QSD and jumps it is useful to consider an analogy. For the Newtonian dynamics of the Earth and the Moon, momentum is continuously transferred between them. But in the impulsive dynamics of billiards, snooker, pool or Newton's balls, momentum is supposed to be transferred instantaneously. A continuous change of momentum can be interpreted as a sequence of small sudden changes or impulses, and this picture is often appropriate for calculations and theory. Conversely, an impulse can be considered as the short-time limit of a continuous transfer of momentum, with the magnitude of the impulse held constant. Either approach is legitimate, but in classical dynamics the continuous theory is usually treated as fundamental and the sudden changes or impulses are considered to be secondary.

Similarly in the unravelling of master equations, either a continuous stochastic equation like a QSD equation, or a sequence of sudden jumps, can be considered as fundamental, and the other derived from it. In this book we show how the jumps can be derived as a limit of QSD. In quantum optics the interaction between an electromagnetic field and a photon counter is continuous. But the response of a photon counter to absorption of a photon is very rapid. It can be considered as instantaneous, and it therefore makes sense to model these measurements of quantum optics by jumps, and that is what is usually done in practice. By contrast, where the response of the environment is naturally considered as continuous, for example in the interaction of an organic molecule with its liquid environment, and in the examples given in chapter 6, QSD is more appropriate. States can be localized both with QSD and with jumps [83]. Localization appears to be more natural for QSD.

Localization is a physical process, but its details are not accessible to experiment. We have no way of deciding from experiment whether it is a continuous diffusion process or a series of stochastic jumps. Localization can be based on continuous stochastic changes or on quantum jumps as fundamental. Either is legitimate, but here, by analogy with Newtonian dynamics, the continuous theory is treated as fundamental, and the jumps as secondary. It should be remembered, as in the case of hard collisions between solid bodies in Newtonian dynamics, that where sudden changes are more important, as in photon counting, jumps are often simpler in theory and in practice. This is important when deciding whether to use QSD in the computer program described in chapter 6.

The application of quantum jump methods to quantum optics was stimulated by Dehmelt's three-level scheme for observing them in single trapped atoms [29, 30]. The method has an extensive literature, including [22, 121, 148, 24, 160, 45, 96, 25, 155, 156].

Garraway and Knight have made comparisons between QSD and quantum jump simulations for dissipative systems in quantum optics [57, 59, 58]. They showed that QSD often localizes where jump simulations do not. With Steinbach they introduced a Monte Carlo method which provides a fourth-order simulation of the master equation [147]. Holland, Marksteiner, Marte and Zoller have shown how to produce localization with jumps by simulating fictional measurements [83].

The time correlation functions and spectra of open quantum systems [90] can be evaluated using quantum jump methods [22, 54] and QSD [128], but it is not simple to find a correlation function for a single QSD trajectory with the correct ensemble mean. This problem has been addressed by Gisin using the Heisenberg picture [71], and by Brun and Gisin who introduce a supplementary state vector trajectory that depends on the original trajectory through an additional QSD equation [20].

4.8 The circuit analogy

The representation of measurement by an Hermitian Lindblad is formal. One of the aims of QSD is to show that this formal treatment of measurement follows from a more detailed treatment of the interaction between the system and the apparatus as its environment, but this depends on the details of the physics of the measuring apparatus, and the choice of boundary. It has not yet been worked out in sufficient detail.

There is an analogy between the state diffusion theory of quantum mechanics and classical circuit theory. An electric circuit includes Hamiltonian elements, with capacitance and inductance, and also non-Hamiltonian elements, such as resistors, which represent the interaction of the circuit with its 'environment', consisting of the internal electronic freedoms of the resistor. This interaction produces stochastic fluctuations in the form of thermal noise, which can be represented by an Itô equation. The resistors are normally treated on the same basis as the Hamiltonian parts of the system.

It is not considered necessary to analyse the detailed physics of every resistor before attempting to solve a problem in circuit theory. But it is important that such an analysis should be possible, for without it there can be no understanding of the physics of the process. A simple example is the aerial treated as a resistor. A more complicated example is the detailed dynamics of electrons in a solid state resistor. This is obviously much more difficult than the representation of a linear resistor by a real impedance in circuit theory.

The simple representation is the analogue of the representation of a quantum measurement by an Hermitian Lindblad, the electron dynamics is the analogue of the treatment of measuring apparatus as the environment of the quantum system and working out the interactions between them in detail. For example, this would require an analysis of the physics of amplifiers and electron multipliers. It has not yet been done in general, but a special case appears in section 6.5.

5
Localization

The localization properties of QSD are important for practical numerical computations and essential for its application to quantum foundations. When a physical variable represented by an Hermitian operator is measured, the state localizes towards a single value of that variable. For more general interactions of the system with its environment, the states of the system tend to localize to wave packets that are confined to small regions of phase space. These properties have been confirmed numerically for a wide variety of simple systems. General properties for wide open systems are derived in this chapter from the QSD equations themselves. Localization produced by QSD is so important that the theory is explained in detail in this chapter, which is more mathematical than the other chapters. If you are only interested in the main results, you can find them in the first and last sections.

5.1 Measurement and classical motion

Stern and Gerlach passed a beam of silver atoms through a magnetic field and detected them with photographic emulsion. The beam was split into two, corresponding to the components of spin in the direction of the magnetic field, as illustrated in figure 5.1.

In a Stern-Gerlach experiment, when the spin of an atom in the initial beam is aligned in a direction perpendicular to the field, the de Broglie wave of the atomic motion is a linear combination of waves of equal amplitude in each beam. The probabilities are equal too, but a single atom is only seen in one of them. The process of detection, which produces a microscopic permanent image of each atom on a photographic plate, appears to localize the atom in one beam or the other. This is sometimes called reduction. The

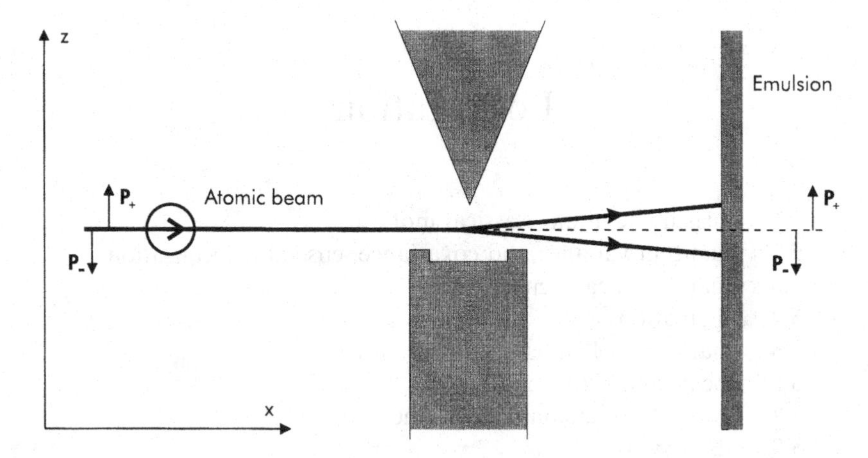

Figure 5.1 Stern-Gerlach apparatus and localization of silver atoms by a photographic emulsion.

Stern-Gerlach experiment is typical of many quantum measurements in the laboratory.

When a gamma ray is emitted by a nucleus, its wave is spread around the nucleus in all directions, but when it is detected by absorption in a fluorescent screen, its effect is seen as a localized flash of light. This is the old problem of wave-particle duality, seen in figure 5.2.

These experiments and others like them cannot be explained by using only Schrödinger's quantum dynamics. For Bohr and von Neumann and almost

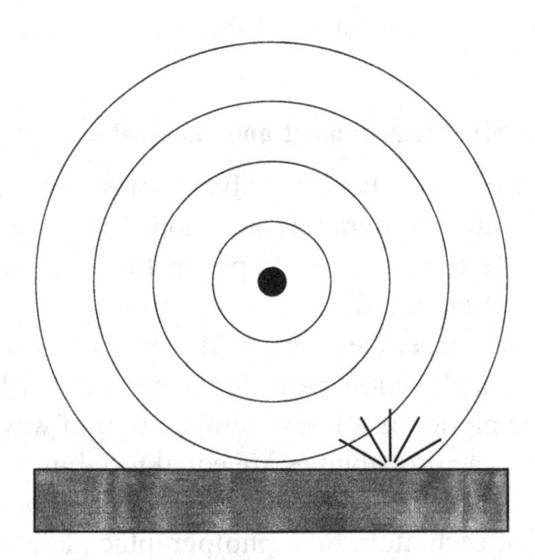

Figure 5.2 Localization of a gamma ray emitted by a nucleus.

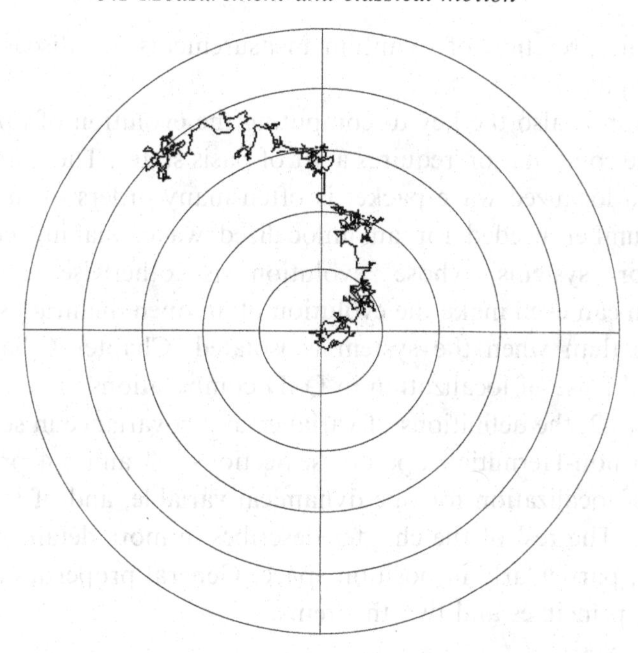

Figure 5.3 The track of a classical Brownian particle and a sketch of the corresponding outgoing de Broglie wave.

all practising quantum physicists since then, the theory of the experiments requires an *interpretation*, which is additional to the dynamics.

As shown by Ehrenfest, classical motion of free particles can be simulated for short periods of time by localized wave packets that are solutions of Schrödinger's equation. Later, according to the equation, these wave packets disperse more or less rapidly into larger and larger regions of space. Yet even small systems like Brownian particles insist on behaving classically for arbitrarily long periods of time, as if there were some process that keeps them localized, as shown in figure 5.3. All classical dynamical variables behave as if there were a process that keeps them localized.

According to QSD there *is* a process that keeps all classical dynamical variables localized, and which also produces the localization in quantum measurements. It is quantum state diffusion. This was the reason for introducing QSD in the first place. When the diffusion takes place as a result of an interaction of the system with its environment, it is immaterial whether that environment happens to include measuring apparatus, or some more general environment. Almost any interaction with an environment produces localization, with different environments producing different kinds of localization. *Measurement is nothing special.* The strengths and weaknesses of QSD and the usual interpretation of quantum measurements are discussed in more detail in chapter 7.

Localization is also the key to computing the evolution of open quantum systems. The computation requires a set of basis states. The number of these needed for a localized wave packet is often many orders of magnitude less than the number needed for an unlocalized wave, making computations possible for systems whose evolution is otherwise uncomputable. Localization can even make the evolution of an open quantum system easier to compute than when the system is isolated. Chapter 6 has a detailed account of the use of localization in QSD computations.

In section 5.2, the definitions of variance and covariance in section 3.7 are extended to non-Hermitian operators. Sections 5.3 and 5.4 present simple examples, of localization for one dynamical variable, and of localization in phase space. The rest of the chapter describes in more detail the physics of localization, particularly in position space. General properties are summarized in four principles and two theorems.

The principles are:

Pri1. Interactions with independent parts of the environment lead to statistically independent quantum state diffusions.

Pri2. *Amplitudes* fluctuate where *probabilities* are conserved: where there are independent environmental interactions in subspaces, ensemble probabilities are conserved, but amplitudes for the individual systems of the ensemble fluctuate, unless the effective rate of interaction of the system with its environment is zero.

Pri3. The above probability conservation and fluctuation together imply localization (with special exceptions).

Pri4. *Position* localization of states in position space follows from locality of interactions. The rest frame of the localization is determined by the environment.

The theorems are the dispersion entropy theorem and the general localization theorem. These appeared in [111].

A classical theory of localization is presented in chapter 9, which deals with the localization of systems with high quantum number, whereas a theory of the localized states is given in section 10.6.

5.2 Quantum variance and covariance, ensemble localization

The definitions of section 3.7 for Hermitian operators are here extended to arbitrary operators. They are natural generalizations of the classical covariance of complex variables in section 2.6, and many of the results of that section carry over to this one.

The *quantum covariance* of two operators for a system in the state $|\psi\rangle$ is

$$\begin{aligned}\sigma(\mathbf{F},\mathbf{G}) &= \langle \mathbf{F}_\Delta^\dagger \mathbf{G}_\Delta\rangle = \langle \mathbf{F}_\Delta^\dagger \mathbf{G}\rangle = \langle \mathbf{F}^\dagger \mathbf{G}_\Delta\rangle \\ &= \langle \mathbf{F}^\dagger \mathbf{G}\rangle - \langle \mathbf{F}\rangle^* \langle \mathbf{G}\rangle. \end{aligned} \tag{5.1}$$

The quantum covariance of $\mathbf{F}$ with itself is just the *quantum variance* of $\mathbf{F}$:

$$\sigma(\mathbf{F},\mathbf{F}) = \langle \mathbf{F}^\dagger \mathbf{F}\rangle - \langle \mathbf{F}\rangle^* \langle \mathbf{F}\rangle = \sigma^2(\mathbf{F}). \tag{5.2}$$

For Hermitian $\mathbf{F}$, $\sigma^2(\mathbf{F}) = (\Delta f)^2$, where Δf is the standard deviation of the measured dynamical variable f.

When $\mathbf{F}$ is non-Hermitian and it does not commute with $\mathbf{F}^\dagger$, the variance $\sigma^2(\mathbf{F})$ is not the same as that of its conjugate $\mathbf{F}^\dagger$, but their average is the sum of the variances $\sigma^2(\mathbf{F}_\mathrm{R})$ and $\sigma^2(\mathbf{F}_\mathrm{I})$. This convention is helpful to the later analysis, but differs from an earlier convention of Caves [23]. In particular, for annihilation and creation operators,

$$\sigma^2(\mathbf{a}) + \tfrac{1}{2} = \sigma^2(\mathbf{a}^\dagger) - \tfrac{1}{2} = \frac{1}{2\hbar}\left(\sigma^2(\mathbf{x}) + \sigma^2(\mathbf{y})\right). \tag{5.3}$$

If $\sigma^2(\mathbf{a}) = 0$, then

$$\sigma^2(\mathbf{x}) + \sigma^2(\mathbf{y}) = \hbar, \tag{5.4}$$

the value for the ground state of an oscillator, or for the more general coherent states, which are minimum indeterminacy wave packets.

The *ensemble localization* $\Lambda(\mathbf{F})$ of an Hermitian or non-Hermitian operator $\mathbf{F}$ is defined as the inverse of the ensemble mean of the quantum variance:

$$\Lambda(\mathbf{F}) = \left(\mathrm{M}\,\sigma^2(\mathbf{F})\right)^{-1}. \tag{5.5}$$

According to this definition, the localization of the annihilation operator $\mathbf{a}$ is infinite for a minimum indeterminacy coherent state.

With these definitions there are strict lower bounds on the rates of localization for wide open systems.

5.3 Quantum measurement

In a Stern-Gerlach experiment, for simplicity on a single silver atom with its spin perpendicular to the field, Schrödinger's equation can be applied to the combined system of atom and photographic plate. The result is a quantum state that is a coherent linear combination of the quantum states of the two possible latent images of the atom on the plate. What is seen is quite different: an image in one place or the other. This is a simple example of the paradox of Schrödinger's cat [133, 132, 154]. It is an example of destructive measurement.

In the case of nondestructive measurement, Schrödinger's equation produces an entanglement between the measured system and the measurer, with non-zero amplitudes for all the possible values of the relevant classical variables of the measurer. Again this is not seen. The classical variable has a more-or-less definite value, the entanglement for an individual run of an experiment is destroyed, but reappears as a correlation between measurer and measured system in an ensemble of many runs.

A *quantum measurement* is an interaction between a quantum system and its environment in which the state of the quantum system significantly influences a classical dynamical variable of the environment. A laboratory measurement is just one example, but there are many others, described in section 3.5. The interaction is not instantaneous, but takes a finite time. Because of the interaction, the system is open, and we can apply QSD. The interaction is always complicated, but for the measurement of a single dynamical variable with operator **F**, the main effect of the interaction on the system can be represented by a single Lindblad operator $\mathbf{L} = c\mathbf{F}$, where c is a positive constant that determines the rate of the measurement.

At first we suppose that the effect of the Hamiltonian evolution of the system can be neglected during the course of the measurement, as in a well-prepared Stern-Gerlach experiment. The system is then wide open. Because **L** is an Hermitian operator, the QSD equation is

$$|d\psi\rangle = -\tfrac{1}{2}\mathbf{L}_\Delta^2|\psi\rangle dt + \mathbf{L}_\Delta|\psi\rangle d\xi, \tag{5.6}$$

where $\mathbf{L}_\Delta$ is the shifted operator defined by (3.22).

For any Hermitian operator **G** that commutes with **L**, the change in the expectation $\langle\mathbf{G}\rangle$ of **G** due to the diffusion of $|\psi\rangle$ in time dt is

$$\begin{aligned} \mathrm{d}\langle \mathbf{G}\rangle &= \langle\psi|\mathbf{G}|\mathrm{d}\psi\rangle + \langle\mathrm{d}\psi|\mathbf{G}|\psi\rangle + \langle\mathrm{d}\psi|\mathbf{G}|\mathrm{d}\psi\rangle \\ &= 2\mathrm{Re}\langle\psi|\mathbf{G}|\mathrm{d}\psi\rangle + \langle\mathrm{d}\psi|\mathbf{G}|\mathrm{d}\psi\rangle \\ &= -\langle \mathbf{G}\mathbf{L}_\Delta^2\rangle \mathrm{d}t + \langle \mathbf{L}_\Delta \mathbf{G}\mathbf{L}_\Delta\rangle \mathrm{d}t + 2\mathrm{Re}(\langle \mathbf{G}\mathbf{L}_\Delta\rangle \mathrm{d}\xi) \\ &= 2\mathrm{Re}(\sigma(\mathbf{G},\mathbf{L})\mathrm{d}\xi) \\ &= 2\sigma(\mathbf{G},\mathbf{L})\mathrm{d}\xi_\mathrm{R}, \end{aligned} \tag{5.7}$$

since $\sigma(\mathbf{G}, \mathbf{L})$ is real. We always use 2Re to represent the addition of the complex conjugate or the Hermitian conjugate to the expression that follows it, sometimes denoted +cc or +HC on the right of the expression in the quantum optics literature.

For

$$\mathbf{L} = c\mathbf{G}, \qquad \text{we have} \qquad \mathrm{d}\langle \mathbf{G}\rangle = 2c\,\sigma^2(\mathbf{G})\mathrm{d}\xi_\mathrm{R}. \tag{5.8}$$

Taking the mean, we get zero, $\mathrm{M}\,\mathrm{d}\langle \mathbf{G}\rangle = 0$. The ensemble mean of the change in expectation is zero, so that the measurement of a dynamical variable does not change the mean of any dynamical variable that commutes with it. In particular it does not change its own mean, which is just as well, as the result would be wrong if it did.

Equation (5.7) is used to obtain the changes in $\langle \mathbf{G}\rangle$ and $\langle \mathbf{G}^2\rangle$ that are needed for the change of the localization $\Lambda(\mathbf{G})$, which is the inverse of the ensemble mean of the variance of $\mathbf{G}$. The change in $\mathrm{M}\,\sigma^2(\mathbf{G})$ is

$$\begin{aligned} \mathrm{M}\,\mathrm{d}\sigma^2(\mathbf{G}) &= \mathrm{M}\left(\mathrm{d}\langle \mathbf{G}^2\rangle - 2\langle \mathbf{G}\rangle \mathrm{d}\langle \mathbf{G}\rangle - (\mathrm{d}\langle \mathbf{G}\rangle)^2\right) \\ &= -2[\sigma(\mathbf{G},\mathbf{L})]^2\mathrm{d}t. \end{aligned} \tag{5.9}$$

When $\mathbf{L} = c\mathbf{G}$, this simplifies to

$$\mathrm{M}\,(\mathrm{d}\sigma^2(\mathbf{G})/\mathrm{d}t) = -2c^2[\sigma^2(\mathbf{G})]^2. \tag{5.10}$$

To go further, look at the whole ensemble over a finite time. This requires an ensemble mean of both sides, giving

$$\frac{\mathrm{d}}{\mathrm{d}t}\mathrm{M}\,\sigma^2(\mathbf{G}) = -2c^2\mathrm{M}\,(\sigma^2(\mathbf{G}))^2 \le -2c^2(\mathrm{M}\,\sigma^2(\mathbf{G}))^2, \tag{5.11}$$

where the inequality follows from the positivity of the ensemble variance of $\sigma^2(\mathbf{G})$ defined by equation (2.8):

$$\Sigma^2(\sigma^2(\mathbf{G})) = \mathrm{M}\left(\sigma^2(\mathbf{G})\right)^2 - \left(\mathrm{M}\,\sigma^2(\mathbf{G})\right)^2 \geq 0. \tag{5.12}$$

So a bound on the change of the ensemble localization of **G**

$$\Lambda(\mathbf{G}) = \left(\mathrm{M}\,\sigma^2(\mathbf{G})\right)^{-1} \tag{5.13}$$

is given by

$$\frac{\mathrm{d}}{\mathrm{d}t}(\Lambda^{-1}) \leq -2c^2\Lambda^{-2},$$

$$\text{so that} \qquad \frac{\mathrm{d}\Lambda}{\mathrm{d}t} \geq 2c^2 \qquad \text{and} \qquad \Lambda(t) \geq \Lambda(0) + 2c^2 t. \tag{5.14}$$

In practice, the value of c is very large when the system becomes entangled with macroscopic variables, as it does during a measurement. Localization then appears to be instantaneous. This is the quantum jump limit of a measurement, giving a physical picture which is consistent with the Copenhagen interpretation.

The above treatment of measurement makes no reference to the physics of the measuring apparatus. It is formal, and one of the aims of QSD is to show that such formal treatments follow from a more detailed theory of the interaction between the measured quantum system and the measuring apparatus as the environment of the system, as discussed in section 4.8. The theory depends on where the division between system and environment is made, and on the detailed physics of the measuring apparatus, including dissipation in amplifiers and resistors. Section 6.5 contains a brief description of such a theory, as applied to the continuous Stern-Gerlach effect by Alber and Steimle.

Quantum measurement is an untypical interaction between system and environment, which produces localization in only one dynamical variable of the system. For most systems, phase space localization is the norm. This has been demonstrated extensively by numerical experiments, and is analysed in the next section and in chapters 9 and 10.

5.4 Dissipation

The simplest types of dissipation are represented by the $\boldsymbol{\sigma}_-$ spin operator and the annihilation operator $\mathbf{a}$, which is used to illustrate phase space localization here. These are non-Hermitian operators, and they do not commute with their conjugates. The theory is not quite so simple as for measurement, and we take the opportunity to derive the expression for the change in the variance $\sigma^2(\mathbf{G})$ of an arbitrary operator $\mathbf{G}$ for a wide open system with any number of arbitrary $\mathbf{L}_j$.

From the QSD equation (4.21) with $\mathbf{H} = 0$, after much cancellation, the change in $\langle\mathbf{F}\rangle$ for arbitrary $\mathbf{F}$ and many Lindblads is

$$\begin{aligned}
\mathrm{d}\langle\mathbf{F}\rangle &= \langle\psi|\mathbf{F}|\mathrm{d}\psi\rangle + \langle\mathrm{d}\psi|\mathbf{F}|\psi\rangle + \langle\mathrm{d}\psi|\mathbf{F}|\mathrm{d}\psi\rangle \\
&= \sum_j \{\langle\mathbf{L}_j^\dagger\mathbf{F}\mathbf{L}_j - \tfrac{1}{2}\mathbf{L}_j^\dagger\mathbf{L}_j\mathbf{F} - \tfrac{1}{2}\mathbf{F}\mathbf{L}_j^\dagger\mathbf{L}_j\rangle \mathrm{d}t \\
&\qquad + \sigma(\mathbf{F},\mathbf{L}_j)\mathrm{d}\xi_j + \sigma(\mathbf{L}_j^\dagger,\mathbf{F})\mathrm{d}\xi_j^*\} \qquad (5.15)\\
&= \sum_j \{\tfrac{1}{2}\langle\mathbf{L}_j^\dagger[\mathbf{F},\mathbf{L}_j] + [\mathbf{L}_j,{}^\dagger\,\mathbf{F}]\mathbf{L}_j\rangle\mathrm{d}t + \sigma(\mathbf{F},\mathbf{L}_j)\mathrm{d}\xi_j + \sigma(\mathbf{L}_j^\dagger,\mathbf{F})\mathrm{d}\xi_j^*\}.
\end{aligned}$$

Putting $\mathbf{F} = \mathbf{G}$ and $\mathbf{F} = \mathbf{G}^\dagger\mathbf{G}$ we find that

$$\begin{aligned}
\mathrm{M}\,\mathrm{d}\sigma^2(\mathbf{G}) &= \mathrm{M}\,\mathrm{d}\langle\mathbf{G}^\dagger\mathbf{G}\rangle - 2\mathrm{M}\,\mathrm{Re}\langle\mathbf{G}^\dagger\rangle\mathrm{d}\langle\mathbf{G}\rangle - \mathrm{M}\,\mathrm{d}\langle\mathbf{G}^\dagger\rangle\mathrm{d}\langle\mathbf{G}\rangle \\
&= \tfrac{1}{2}\,\mathrm{M}\,\langle\mathbf{L}_j^\dagger[\mathbf{G}^\dagger\mathbf{G},\mathbf{L}_j] + [\mathbf{L}_j^\dagger,\mathbf{G}^\dagger\mathbf{G}]\mathbf{L}_j\rangle\mathrm{d}i \qquad \text{(commutation)} \\
&\quad - \mathrm{M}\,\mathrm{Re}(\langle\mathbf{F}^\dagger\rangle\langle\mathbf{L}_j^\dagger[\mathbf{G},\mathbf{L}_j] + [\mathbf{L}_j^\dagger,\mathbf{G}]\mathbf{L}_j\rangle)\mathrm{d}t \qquad \text{(commutation)} \\
&\quad - \mathrm{M}\,(|\sigma(\mathbf{G},\mathbf{L}_j)|^2 + |\sigma(\mathbf{L}_j^\dagger,\mathbf{G})|^2)\mathrm{d}t \qquad \text{(covariance).}
\end{aligned} \qquad (5.16)$$

The commutation terms can have either sign, but the covariance term must always be negative. So a Lindblad $\mathbf{L}$ localizes an operator $\mathbf{G}$ for which the commutators vanish, unless the covariances are both zero. But if the operators do not commute, then the commutators can produce delocalization. This is what happens for conjugate variables, so that they satisfy Heisenberg indeterminacy despite the localization. However, in the classical theory of chapters 9 and 10, the commutation terms are negligible, and even conjugate variables are localized.

Now apply the general result (5.16) to localization due to dissipation, when there is a single Lindblad proportional to the annihilation operator:

$$\mathbf{L} = \gamma^{\frac{1}{2}}\mathbf{a}, \qquad \mathbf{G} = \mathbf{a}, \qquad (5.17)$$

where γ is a rate. The commutation relation $[\mathbf{a},\mathbf{a}^\dagger] = 1$ gives

$$\mathrm{M}\,\mathrm{d}\sigma^2(\mathbf{a}) = -\gamma\mathrm{M}\left[\sigma^2(\mathbf{a}) + \left(\sigma^2(\mathbf{a})\right)^2 + |\sigma(\mathbf{a}^\dagger,\mathbf{a})|^2\right]\mathrm{d}t, \qquad (5.18)$$

which is always negative unless both $\sigma^2(\mathbf{a})$ and $\sigma(\mathbf{a}^\dagger, \mathbf{a})$ are zero. Using the inequality (5.12), dropping the last term in (5.18) above, and putting $\Lambda = 1/\mathrm{M}\,\sigma^2(\mathbf{a})$ gives

$$d(1/\Lambda) \leq -\gamma(1/\Lambda + 1/\Lambda^2)dt \qquad \text{or} \qquad d\Lambda \geq \gamma(\Lambda + 1). \tag{5.19}$$

To get the boundary of the inequality over a finite time, we replace it by an equality and integrate, to get

$$\Lambda(t) \geq (\Lambda(0) + 1)e^{\gamma t} - 1. \tag{5.20}$$

The localization increases exponentially at a rate given by γ, which is also the rate of dissipation. In the mean, the quantum wave localizes exponentially at this rate to a coherent state, which is a minimum indeterminacy wave packet.

5.5 Channels and statistical properties

Channels help us to understand the physics of measurement, and to analyse the statistical properties of quantum states.

The space of states $|\psi\rangle$ of a quantum system may be divided or *partitioned* into orthogonal subspaces or *channels*. The channels are labelled by the complete set of projectors $\mathbf{P}_k$ that project onto them, where

$$\mathbf{P}_k\mathbf{P}_\ell = \mathbf{P}_\ell\mathbf{P}_k = \delta_{k\ell}\mathbf{P}_\ell, \qquad \sum_k \mathbf{P}_k = \mathrm{I}. \tag{5.21}$$

The partitions are often spatially separated. In that case the projectors project onto mutually exclusive regions of position space, and the channels can be thought of as regions of ordinary position space. For the Stern-Gerlach experiment, position space is divided into two parts, each containing a beam of atoms with a + or − spin. More generally, channels correspond to the eigenspaces of Hermitian operators. When these operators represent dynamical variables, then the channels label values or ranges of those variables. A channel may contain one state, an infinite continuum of states, or anything in between. We can ignore transitions between the internal states of each channel, and consider only the transitions between the channels, which are represented by relatively simple operators.

A complicated interaction that produces localization in eigenspaces of a dynamical variable with Hermitian operator $\mathbf{G}$ can be represented by an Hermitian Lindblad $c\mathbf{G}$. If the operators are chosen well, then the complicated physics of more general interactions can be represented adequately, though approximately, by simple models with a few Lindblads, not necessarily Hermitian.

One simplification is that each channel, consisting of any number of states, can be represented by a one-state channel. The transitions between channels are then represented by relatively simple operators. For example annihilation and downward transition operators can represent dissipation by the environment, and a complicated interaction that produces localization in eigenspaces of a dynamical variable can be represented by an Hermitian environment operator $c\mathbf{G}$. If the operators are chosen well, then localization of quite complicated systems can be represented adequately, though approximately, by these simple models.

5.6 Localization theorems

This section states two theorems about localization in different channels: the dispersion entropy theorem and the general localization theorem. The latter is proved here, and the former in the next section. Other properties of localization are derived in chapters 9 and 10 on classical and semiclassical theory.

Suppose there are two or more channels defined by a complete set of independent projectors $\mathbf{P}_k$.

A *block diagonal operator* has no nonzero off-diagonal matrix elements coupling the channels, so the nonzero elements lie in square blocks along the diagonal, each block representing a channel, as in figure 5.4. For the

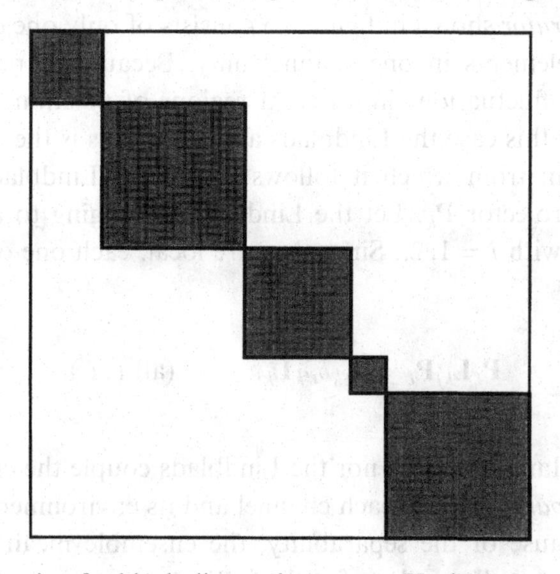

Figure 5.4 The matrix of a block diagonal operator. Unshaded areas are zero.

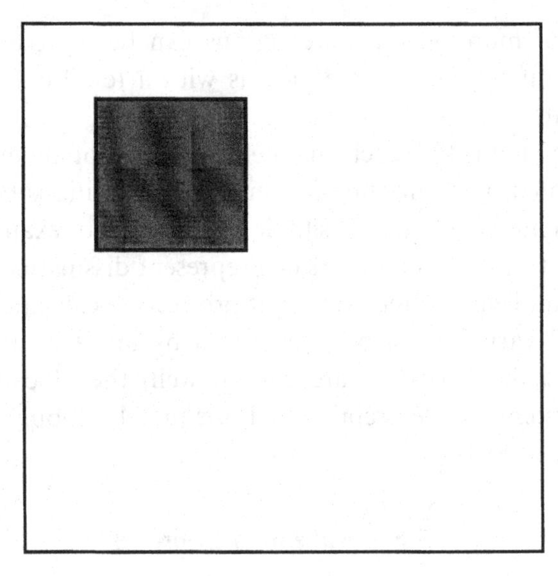

Figure 5.5 The matrix of a local operator. Unshaded areas are zero.

dispersion entropy theorem, we assume that the Hamiltonian interaction between channels is negligible compared with the localization rate. Consequently the Hamiltonian is block diagonal and for every channel projector $\mathbf{P}_k$, we have $[\mathbf{H}, \mathbf{P}_k] = 0$.

The *local operator* shown in figure 5.5 consists of only one diagonal block, with nonzero elements in one channel only. Because of principle Pri1 of section 5.1, the fluctuations in different regions of position space are independent. So for this case the Lindblads are local. This is the other condition for the theorem, from which it follows that every Lindblad belongs to a channel with projector $\mathbf{P}_k$. Let the Lindblads belonging to a particular $\mathbf{P}_k$ be labelled $\mathbf{L}_{kj}$ with $j = 1, 2,$. Since they are local, each one of them satisfies the condition

$$\mathbf{P}_\ell \mathbf{L}_{kj} \mathbf{P}_{\ell'} = \delta_{\ell k} \delta_{\ell' k} \mathbf{L}_{kj} \qquad (\text{all } \ell, \ell'). \tag{5.22}$$

Neither the Hamiltonian $\mathbf{H}$ nor the Lindblads couple the channels. So the channels are *separable*, since each channel and its environment operate independently. Because of the separability, the ensemble mean $\mathbf{M}\langle\mathbf{P}_k\rangle$ is conserved. Later we shall see that, for the individual systems of the ensemble, the *weight* W_k, which is the conditional probability $\langle \mathbf{P}_k \rangle$ for being in channel k, is not conserved. This is the nonlocal property of the localization process: despite the lack of interaction between the channels, which are often separated in space, the weights change.

A normalized state vector $|\psi\rangle$ can be represented as a linear combination of normalized channel state vectors

$$|\psi\rangle = \sum_k a_k |\psi_k\rangle \qquad (\langle \psi_k | \psi_k \rangle = 1), \tag{5.23}$$

where

$$|a_k|^2 = \langle \mathbf{P}_k \rangle = W_k \tag{5.24}$$

is the weight of the state $|\psi\rangle$ in the channel $\mathbf{P}_k$. The expectations of the Lindblads are

$$\langle \mathbf{L}_{kj} \rangle = \langle \psi | \mathbf{L}_{kj} | \psi \rangle = \langle \psi_k | \mathbf{L}_{kj} | \psi_k \rangle, \tag{5.25}$$

where $\mathbf{L}_{kj}$ is the jth Lindblad belonging to channel $\mathbf{P}_k$. The *effective interaction rate* in channel k is then defined as

$$R_k = \sum_j |\langle \mathbf{L}_{kj} \rangle|^2, \tag{5.26}$$

which has the dimensions of inverse time.

The dispersion or delocalization for state $|\psi\rangle$ and partition $\mathbf{P}_k$ is measured by the *quantum dispersion entropy* Q:

$$Q = Q(|\psi\rangle, \mathbf{P}_k) = -\sum_k W_k \ln W_k. \tag{5.27}$$

Evidently, the smaller the entropy, the greater the localization among the channels.

The *dispersion entropy theorem* states that the mean rate of decrease of quantum dispersion entropy for the partition is minus a weighted sum over effective interaction rates:

$$\frac{d}{dt}(\mathrm{M}\,Q) = -\sum_k \frac{1 - W_k}{W_k} R_k$$
$$= -\sum_k W_k(1 - W_k)(R_k/W_k^2) \le 0, \tag{5.28}$$

where R_k/W_k^2 are nonsingular. For only two complementary channels,

$$\frac{d}{dt}(\mathrm{M}\,Q) = -\left(\frac{W_2 R_1}{W_1} + \frac{W_1 R_2}{W_2}\right)$$
$$= -W_1 W_2(R_1/W_1^2 + R_2/W_2^2) \le 0. \tag{5.29}$$

The mean quantum dispersion entropy always decreases, so the localization always increases, unless the effective interaction rate is zero, or the localization is complete. Section 5.8 discusses this result. The proof is given in section 5.7.

The second theorem of this section is the general localization theorem, for which the conditions are weaker. The Hamiltonian can couple the channels and the Lindblads $\mathbf{L}_j$ are block diagonal in the channel subspaces, so that, for all $\mathbf{P}_k$ and $\mathbf{L}_j$,

$$\mathbf{P}_k\mathbf{L}_j = \mathbf{L}_j\mathbf{P}_k = \mathbf{P}_k\mathbf{L}_j\mathbf{P}_k. \tag{5.30}$$

The *general localization theorem* for an initial pure state then says that for each projector $\mathbf{P}_k$ the change in the ensemble mean of the variance is

$$\mathrm{dM}\,\sigma^2(\mathbf{P}_k) = \mathrm{M}\,(1 - 2\langle\mathbf{P}_k\rangle)\langle -i\hbar^{-1}[\mathbf{H}, \mathbf{P}_k]\rangle \mathrm{d}t$$
$$- \mathrm{M}\,(|\sigma(\mathbf{P}_k, \mathbf{L}_j)|^2 + |\sigma(\mathbf{L}_j^\dagger, \mathbf{P}_k)|^2)\mathrm{d}t. \tag{5.31}$$

The Hamiltonian term follows from (3.2) and $\mathbf{P}_k^2 = \mathbf{P}_k$. The remaining terms follow from equation (5.16) with $\mathbf{G} = \mathbf{P}_k$, where all sums but the last are zero because $\mathbf{P}_k$ commutes with $\mathbf{L}_j$. If the Hamiltonian, like the Lindblads, is block diagonal in the subspaces of $\mathbf{P}_k$, then the first term is zero, and the system localizes in the space of each $\mathbf{P}_k$:

$$\frac{d}{dt}\mathrm{M}\,\sigma^2(\mathbf{P}_k) \le 0, \tag{5.32}$$

with equality when

$$\langle\mathbf{P}_k\mathbf{L}_j\rangle - \langle\mathbf{P}_k\rangle\langle\mathbf{L}_j\rangle = \sigma(\mathbf{P}_k, \mathbf{L}_j) = 0 \qquad (\text{for all } k, j), \tag{5.33}$$

which is a very special case.

When the Hamiltonian is block diagonal, the expectation of each projector tends to zero or one, and the system can only localize into one of the channels.

When a de Broglie wave enters a medium such as a solid, liquid or a dense gas, the Hamiltonian then represents the diffusion of the particle within the medium. The change in position as a result of the diffusion is on a microscopic scale during the localization, so even when the Hamiltonian is not zero, its effects are negligible, and a particle which interacts with material in a number of regions of space localizes into just one of them. An example is the localization of a silver atom by the photographic emulsion in the Stern-Gerlach experiment.

5.7 Proof of the dispersion entropy theorem

For the change in the dispersion entropy (5.27) we need to evaluate $\mathrm{d}(x \ln x)$ to second order in $\mathrm{d}x$, which is

$$\mathrm{d}(x \ln x) = (1 + \ln x)\mathrm{d}x + (\mathrm{d}x)^2/(2x). \tag{5.34}$$

Since $\mathrm{M}\,\mathrm{d}W_k = 0$, only the quadratic term contributes to the change in the entropy

$$\mathrm{M}\,\mathrm{d}(W_k \ln W_k) = \mathrm{M}\,(\mathrm{d}W_k)^2/(2W_k) \tag{5.35}$$

and so

$$\mathrm{M}\,\mathrm{d}Q = -\sum_k \mathrm{M}\,(\mathrm{d}W_k)^2/(2W_k). \tag{5.36}$$

Now use equation (5.15) for the change in $\langle \mathbf{P}_k \rangle$. Since the Lindblads are local, they are also block diagonal, and the commutators are zero. The quantum covariances are

$$\left.\begin{aligned} \sigma(\mathbf{P}_k, \mathbf{L}_{kj}) &= (1 - W_k)\langle \mathbf{L}_{kj} \rangle = \sigma(\mathbf{L}_{kj}^\dagger, \mathbf{P}_k)^*, \\ \sigma(\mathbf{P}_k, \mathbf{L}_{k'j}) &= -W_k \langle \mathbf{L}_{k'j} \rangle = \sigma(\mathbf{L}_{k'j}^\dagger, \mathbf{P}_k)^* \qquad (k' \neq k), \end{aligned}\right\} \tag{5.37}$$

so

$$(\mathrm{d}W_k)^2 = (\mathrm{d}\langle \mathbf{P}_k\rangle)^2 = 2(1-W_k)^2\sum_j |\langle \mathbf{L}_{kj}\rangle|^2 \mathrm{d}t + 2W_k^2 \sum_{k'\neq k}\sum_j |\langle \mathbf{L}_{k'j}\rangle|^2 \mathrm{d}t$$
$$= 2(1-2W_k)R_k\mathrm{d}t + 2W_k^2\sum_{k'} R_{k'}\mathrm{d}t. \tag{5.38}$$

Therefore, since $\sum_k W_k = 1$,

$$\frac{\mathrm{d}}{\mathrm{d}t}\mathrm{M}\,Q = -\mathrm{M}\sum_k \frac{1-2W_k}{W_k}R_k - \mathrm{M}\sum_k W_k \sum_{k'} R_{k'}. - \sum_{j,k'} |\langle \mathbf{L}_{k'j}\rangle|^2$$
$$\leqslant -\sum_k \frac{1-W_k}{W_k}R_k, \tag{5.39}$$

which proves the dispersion entropy theorem.

5.8 Discussion

First consider the four principles of section 5.1.

Independent parts of the environment of a system usually act independently, and so are represented by different sets of environment Lindblads $\mathbf{L}_j$, and statistically independent fluctuations $\mathrm{d}\xi_j$. This is the principle Pri1, which can be applied to the diffusion of a particle by a gas, or the absorption of a particle by a screen.

Localization needs *interaction* between system and environment, producing entanglement between them. One-way action of system on environment is not enough (section 4.6). Consequently the scattering of a particle by a fixed scattering centre has zero effective interaction rate, since there is zero recoil, and if the scatterer is very heavy relative to the system, the effective interaction rate and resultant entanglement are very small. These are examples of action of the environment on the system, and not of interaction between environment and system. They can be represented exactly or approximately by an additional fluctuating Hamiltonian, not by Lindblad environment operators, and they do not produce localization. The same goes for the focussing and diffraction of photons by nonabsorptive optical apparatus, or the electrons in an electron microscope, where the recoil of the apparatus is negligible.

The principle Pri2 follows directly from the state diffusion equation (4.21). Amplitude fluctuations always occur unless $|\psi\rangle$ is a simultaneous eigenstate of all the Lindblads, which happens when

$$\mathbf{L}_j|\psi\rangle = \langle\mathbf{L}_j\rangle|\psi\rangle \qquad \text{(all } j) \qquad (5.40)$$

and which is clearly exceptional.

If the Hamiltonian $\mathbf{H}$ and the Lindblads $\mathbf{L}_j$ are block diagonal, so that they do not couple the channels significantly, the ensemble probability of being in any of the channels is conserved. But for the state $|\psi\rangle$ of an individual system of the ensemble, the magnitude and phase of the channel components $\mathbf{P}_k|\psi\rangle$ all fluctuate, unless (5.40) is satisfied. There is more detail in the rest of this chapter.

Pri3 is a consequence of the localization theorems.

Pri4 is about the special case of position localization. In that case the projectors $\mathbf{P}_k$ project onto mutually exclusive regions of position space, and, provided that these regions are not too small, the Hamiltonian coupling between the different parts of the environment in the different regions can be neglected. Further, it is a consequence of the approximate locality of interactions that the environment of the region $\mathbf{P}_k$ has no effect on the system when the system is in a different region $\mathbf{P}_{k'}$, and so the conditions of the dispersion entropy theorem are easily and very generally satisfied for the position variable.

The permanence of position localization depends on the standard of rest, or rest frame. In any frame that moves significantly with respect to the physical environment, localization is not permanent, because the system will necessarily move from one region to another, spoiling the position localization. So the environment provides the reference frame for position localization.

We have shown how the quantum state diffusion equations of an open system represent explicitly the process of localization in a subspace or channel of the state space. The localization due to the environment increases when the ensemble mean of the quantum variance or the quantum dispersion entropy of the channel projectors decreases, and the two theorems demonstrate that this always happens unless the change is zero. Explicit formulae for the rates of change are obtained in terms of effective interaction rates. Localization is characteristic of the interaction of a system with its environment, whether or not that environment contains measuring apparatus or an observer.

It may seem remarkable that the mean quantum dispersion entropy should behave contrarily to every other physical entropy under the conditions given here. But this is just a consequence of the remarkable properties of quantum mechanics, in which wave properties of a system, which require it to be extended, are followed by particle properties, in which it is localized, as for example in the two-slits experiment. The dispersion entropy theorem is a mathematical expression of these properties. The reduction of mean dispersion entropy is consistent with the increase of thermodynamic entropy because ensemble entropies increase as a result of the diffusion of pure quantum states, and this more than compensates for the decrease in the mean over the quantum dispersion entropies of the individual pure states of the ensemble.

The Hamiltonian term can increase the mean quantum dispersion entropy, and often does, so in general this entropy can either increase or decrease. This contrasts with the entropy theorem of the classical theory in chapter 9.

In the QSD picture the localization process appears explicitly, whereas in the usual picture the density operator gives an average over the localized states. The reduction of its off-diagonal elements due to the environment has been associated with the localization process before, as described in detail by Zeh and his collaborators, and also by Omnès. This view is further discussed in chapter 7.

Depending on the nature of the system and its interaction with the environment, there may be localization in many different dynamical variables, but localization in position is particularly important because interactions are localized in position space, but not in momentum or other dynamical variables.

For compound systems with many particles, the interaction is not localized in the configuration space of the system, but the interaction of each distinguishable particle with the environment is localized in position space, and the overall effect is to localize each particle in position space, which is equivalent to localizing the whole system in configuration space.

6

Numerical methods and examples

After a brief summary of the methods that can be used to study open quantum systems, the formulation and numerical methods of QSD are described, with particular emphasis on the moving basis. This is followed by summaries of various applications which illustrate the use of the program and show something of the variety of problems that can be tackled using QSD. The final section explains how to obtain and use a general-purpose QSD computer program, using the computations needed to prepare some of the figures in chapter 4 as examples. Key references are [127, 130].

6.1 Methods

The problems are to simulate the evolution of an open quantum system, and to compute the properties of an ensemble of open quantum systems. Numerical solutions are needed for all but the simplest systems, because analytic solutions cannot be found. All the methods described here depend for their formulation on the Lindblads. Fortunately, past research on master equations has produced a large stock of Lindblads for a variety of processes and systems, which can now be used for QSD. The same past experience is very valuable for the choice of boundary between system and environment, described in section 3.6.

There are several numerical methods (NM) for solving these problems, including quantum state diffusion:

NM1. *Master equation.* Direct stepwise integration for the ensemble. This is expensive in computer resources, but provides the most accurate results when it is possible to use it. It does not simulate the evolution of individual systems.

The remaining methods are based on the unravelling of the master equation in different ways into an ensemble of pure states. They are sometimes known as *quantum trajectory* or *Monte Carlo simulation* methods.

NM2. *Quantum jump methods.* These involve continuous integration of the drift term, including the Schrödinger term, interrupted by discrete quantum jumps representing the stochastic part of the interaction with the environment, usually considered as a measurement process. These have been the most widely used in quantum optics, where they were developed. They are at their best when the main environmental effect is a measurement involving particle counting, normally photon counting, or a process which is easy to simulate in this way.

NM3. *Quantum state diffusion, QSD.* This is often preferable when the environment changes continuously instead of by discrete jumps, and when its influence on the system is well represented by a known master equation. QSD is particularly efficient when there is strong localization.

NM4. *The moving basis.* This is also known as *mixed classical-quantum* representation. It is not really a different method, but a supplement to the quantum trajectory methods, in which the basis states are made to change with time. It makes a big improvement in the efficiency of these methods when there is significant localization in phase space and is very effective as the motion approaches the classical limit. It is of little help for the master equation.

Quantum jump methods now have an extensive literature, some of it referenced in the introduction and in section 4.7, so this chapter concentrates on QSD, using an easily available computer library prepared by Schack and Brun [130], which we will call the QSD library. The QSD library can also be used for jump methods and for mixtures of diffusion and jumps, where appropriate.

The solution of a time-dependent Schrödinger equation for an isolated system, or a QSD equation for a single open system of an ensemble, is the trajectory of a pure quantum state in Hilbert space. The computer representation of the state $|\psi(t)\rangle$ at a given time t requires a finite basis of (usually) orthonormal states $|u_n\rangle$ such that

$$
\begin{aligned}
|\psi\rangle &= \sum_{n=1}^{n_\mathrm{x}} |u_n\rangle\langle u_n|\psi\rangle \\
&= \sum_{n=1}^{n_\mathrm{x}} c_n|u_n\rangle = \sum_{n=1}^{n_\mathrm{x}} (c_{n\mathrm{R}} + ic_{n\mathrm{I}})|u_n\rangle.
\end{aligned}
\tag{6.1}
$$

The number of basis states n_x may be the dimension of the state space, for example when the system consists of a finite number of spins. But usually it is chosen as an approximation to a state space of infinite dimension, for example when the Hamiltonian and Lindblads include positions, momenta, or field amplitudes with infinite spectra. The representation of the state is then approximate too. The state is represented in the computer by the $2n_x$ real numbers c_{nR}, c_{nI}.

The evolution of the system over a time δt is simulated by matrix multiplications of state vectors, where the matrices represent the Hamiltonian and Lindblad operators. In general they are $n_x \times n_x$ matrices, but whenever possible the representation is chosen so that the matrices are sparse, so that the number of non-zero or significant elements is proportional to n_x. This reduces the time of computation by a factor of order n_x, but puts a severe constraint on the choice of representation.

Computations are limited both in space (computer store or memory) and in time. Despite enormous advances in computer technology, these limitations are still important. For a single trajectory, only a few operations are performed on each number $c_n(t)$ for a given time t. The time of computation for a single quantum trajectory depends roughly linearly on the space used in the computer store, and so we only need consider the space. Even when the Hamiltonian and Lindblad operators are represented by sparse matrices, the density operator rarely is. The store for a state vector needs to hold about $2n_x$ real numbers, but the store for a density operator needs to hold n_x^2 real numbers. So the solution of the master equation almost always needs much more space and time than a single run for a trajectory.

In practice we find that even a few runs of the QSD equations can tell us a lot about an open system, just like a few runs of a laboratory experiment. This analogy still holds when the system is far away from any laboratory, as for an atom in the atmosphere of a star. In fact, running QSD on a computer is very similar to running a laboratory experiment. In such situations QSD equations gain over master equations by a factor of order n_x in computer space and in time. In practice it is usually found that the space limitation is more severe.

However, for many purposes, ensemble means are the significant quantities, in which case QSD requires many runs with different fluctuations in order to build up statistics. In this mode QSD is a Monte Carlo method, with all its advantages and disadvantages. It is like a computer model of a laboratory experiment in which the number of runs is sufficient to build up 'good statistics'. A computer experiment, like a laboratory experiment, needs careful design and planning, and the runs take a lot of time. Under these circumstances numerical solution of the master equation is always more accurate and sometimes quicker, provided it is possible to use it. But very often it is not possible, because there is not enough space in the computer. Simulation methods like QSD or quantum jumps are then the only ones available.

6.2 Localization and the moving basis

For systems with field modes or moving particles, localization often gives us a method of further improving the efficiency of computation by a large factor. Suppose the system has operators with corresponding dynamical variables of d freedoms and a $2d$-dimensional phase space with vector coordinate and momentum (q, p).

The volume of a region of phase space has dimensions of $(\text{action})^d$, which can be measured in Planck units of $(2\pi\hbar)^d$, the volume of a phase space *Planck cell.* This measures the number of well-chosen basis states needed to represent a state vector which is confined within the region [109]. We need a measure of the approximate effective volume V of phase space occupied by a pure state $|\psi\rangle$. This is provided by taking the overlap with standard shifted wave packets, coherent states $u_{q_0p_0}$, for which the expectations of their d-dimensional coordinate and momentum are (q_0, p_0). The effective volume V is defined as the volume for which the overlap is greater than some prescribed small value ϵ', as illustrated in figure 6.1. The circles of area $2\pi\hbar$ in the figure represent coherent states $|\alpha\rangle$ with overlap $|\langle\psi|\alpha\rangle|^2 = \epsilon'^2$. See section 10.5 for more about coherent states and their representation.

For typical Hamiltonians and Lindblads, an approximate position representation, or alternatively the number states of a harmonic oscillator, give suitable bases, in which the operators representing dynamical variables are

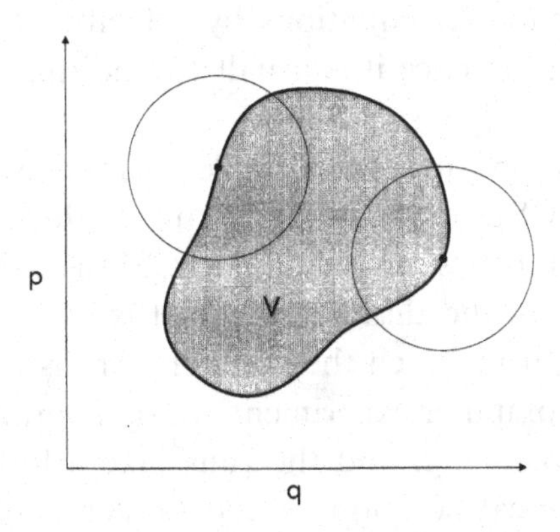

Figure 6.1 The effective volume V in phase space.

represented by sparse matrices. Since most applications have been in quantum optics, where the field modes are oscillators, and nonlinearities are usually small, the oscillator basis is used here and in the computer library of section 6.7.

The minimum number of states required to represent $|\psi\rangle$ to a given precision is then given approximately by the number of Planck cells in V. As the state of a system localizes in phase space, fewer basis states are needed. If the localization is sufficiently strong, V may be reduced to a few Planck volumes, with a corresponding small number of basis states. This small volume moves around in the phase space along an approximately classical trajectory, so, to take advantage of the localization, the basis must follow the trajectory. The detailed theory is given in chapter 10. This is the moving basis [146, 127]. The volume of phase space occupied by the ensemble is almost always much bigger than the volume occupied by the pure states, and so localization rarely works for the master equation. This is shown particularly clearly in the example of the chaotic motion of the Duffing oscillator in the next section.

In practice, the current values of the expectations $\langle \mathbf{q} \rangle$, $\langle \mathbf{p} \rangle$, which define the centroid of the wave packet, are used as the centre of an oscillator basis, with Fock states $|n\rangle$ obtained by shifting from a standard oscillator basis centred at the conventional origin of the phase space. States are included up to some maximum value n_{x}. With an appropriate choice of conjugate coordinates and momenta, these cover a circular or $2d$-dimensional hyperspherical volume of phase space up to some precision which is determined by the program. To maintain precision, the hypersphere must include the volume V, as illustrated in figure 6.2.

6.3 Dissipative quantum chaos

Now for specific applications, using QSD with a moving basis.

Classical chaos in dissipative and Hamiltonian systems has been widely studied by numerical integration of the equations of motion for a sample of trajectories. The quantum chaos of Hamiltonian systems has also been well studied, mainly by analysing spectra. The nature of the classical limit for Hamiltonian systems is a subtle one.

By comparison, quantum dissipative systems have been neglected. This is partly because the density operator does not represent individual systems. Quantum trajectories represent the evolution of individual systems. They are the quantum analogue of classical trajectories, and localization makes the classical limit of QSD relatively straightforward in principle. In practice it is a numerical challenge, because, as the ratios of the classical actions to Planck's

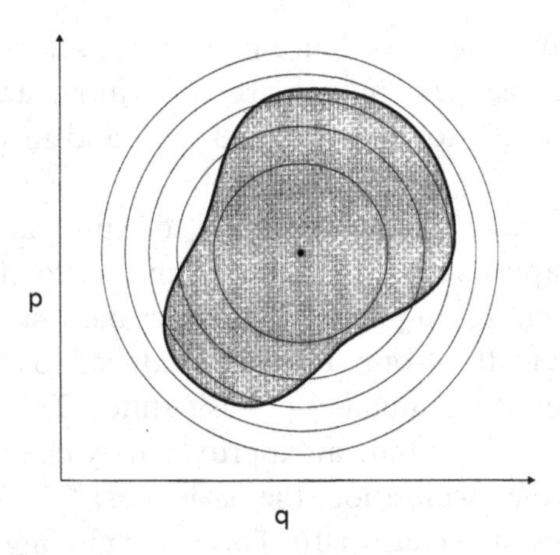

Figure 6.2 Shifted oscillator eigenstates and the volume V.

constant increase, so does the effective size of any fixed basis used to represent the quantum states. The moving basis is ideal for this situation. It becomes relatively more and more efficient as the classical limit is approached.

Spiller and Ralph [140] were the first to study the emergence of chaos in open quantum systems, in a dissipative quantum map. Here we discuss the work of Brun and his collaborators [21, 77], who used the QSD program library of section 6.7 to look at the classical limit of the quantum Duffing oscillator.

The classical Duffing oscillator is a damped, driven, one-freedom system with the double well potential

$$V(x) = \frac{x^4}{4} - \frac{x^2}{2}, \tag{6.2}$$

a driving force of the form $g\cos(t)$ and a damping $-2\Gamma\dot{x}$ proportional to the velocity. The classical equations of motion are

$$\frac{\mathrm{d}^2 x}{dt^2} + 2\Gamma\frac{\mathrm{d}x}{dt} + x^3 - x = g\cos(t). \tag{6.3}$$

Because of the time dependence, the solutions lie in a phase space of three dimensions, and the evolution is conveniently represented by the intersection of the trajectories with a Poincaré surface of section. It has a chaotic regime with a strange attractor.

For the master equation and for QSD, the Hamiltonian is

$$\mathbf{H} = \frac{\mathbf{p}^2}{2} + \frac{\beta^2 \mathbf{q}^4}{4} - \frac{\mathbf{q}^2}{2} + \frac{g}{\beta}\cos(t)\,\mathbf{q} + \sqrt{\Gamma}(\mathbf{qp} + \mathbf{pq}) \tag{6.4}$$

and there is one Lindblad, which is

$$\mathbf{L} = 2\sqrt{\Gamma}\mathbf{a} = \sqrt{2\Gamma}(\mathbf{q} + i\mathbf{p}), \tag{6.5}$$

where $\mathbf{a}$ is the annihilation operator. We put $\hbar = 1$.

The constant β is a scaling factor, which increases the scale of the classical actions as $\beta \to 0$, whilst keeping $\hbar$ constant, so this corresponds to the classical limit.

The intersections of the QSD phase space trajectory $\langle \mathbf{q}(t) \rangle$, $\langle \mathbf{p}(t) \rangle$ with the surface of section are illustrated in figure 6.3, for $\Gamma = 0.125$, $g = 0.3$ and with four different values of β. The quantum system with $\beta = 1$ can be represented by 10 basis states. So it can easily be solved using any method, including the master equation. The chaotic behaviour is entirely masked by quantum fluctuations of the centroid. QSD with a moving basis was used to investigate the transition from this quantum regime to the classical limit, which is a far more demanding computational problem. For $\beta = 0.01$ the quantum surface of section is indistinguishable in form from the classical plot, and the intermediate values of β clearly show the transition from quantum to classical dynamics.

The final effective value for $\hbar$ for (d) was scaled down by a factor of 10 000 compared with its initial value for (a). The density operator would have needed more than 10^9 real numbers, and a fixed basis about 50 000 real numbers. For this near approach to the classical limit, QSD with a moving basis needed about 15 basis states.

The same system has been studied in the adiabatic or rotating wave approximation by Rigo, Alber, Mota-Furtado and O'Mahony [124]. They use QSD to show that an individual quantum system can exhibit bistability, going over smoothly to the classical limit, with a mean transition time for the switching between the two equilibrium points due to quantum fluctuations. This resolves a controversy concerning this limit.

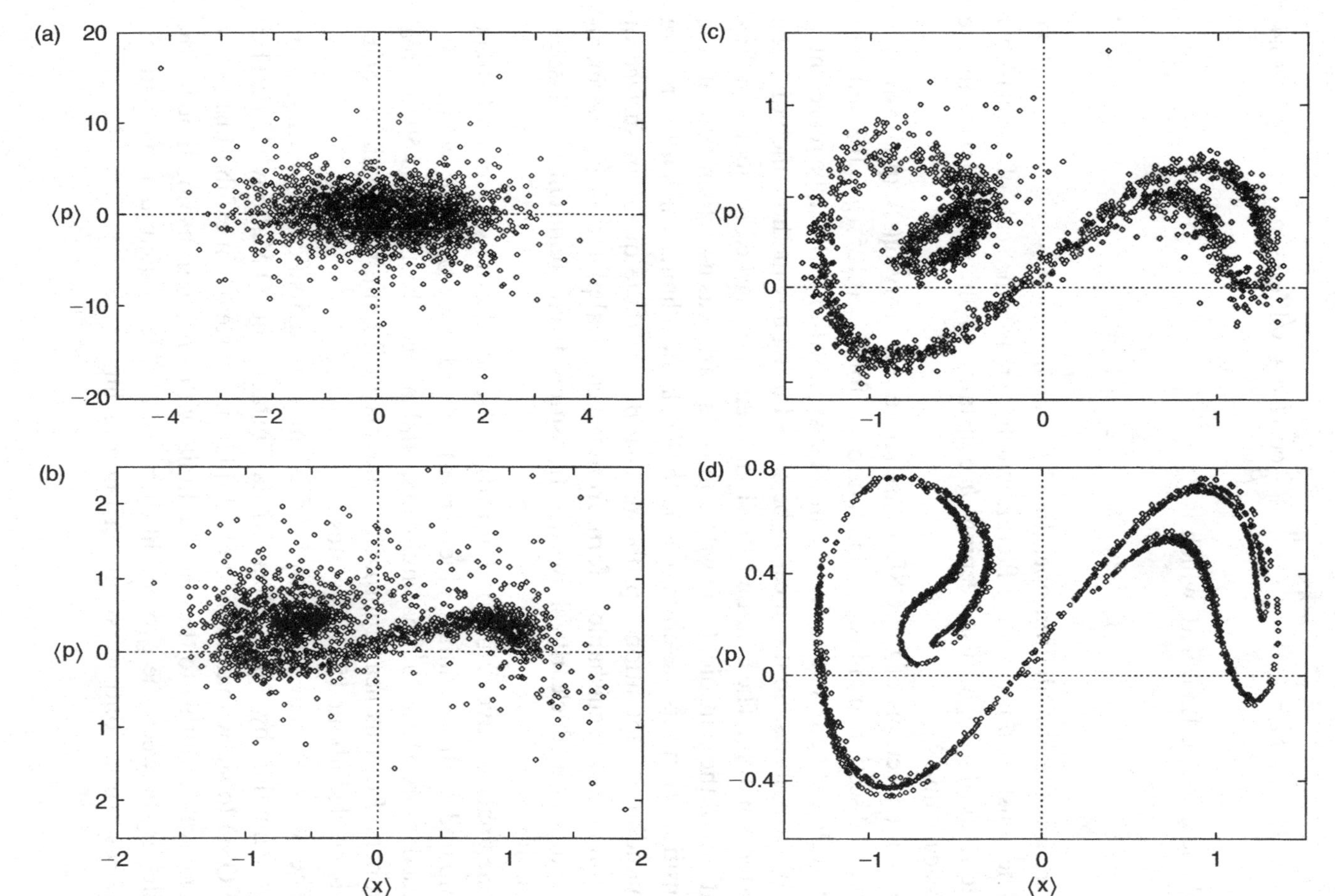

Figure 6.3 The Poincaré surface of section for a single QSD trajectory of a forced damped Duffing oscillator in the chaotic regime with $\Gamma = 0.125$, $g = 0.3$ and four scalings: (a) $\beta = 1.0$, (b) $\beta = 0.25$, (c) $\beta = 0.1$ and (d) $\beta = 0.01$.

6.4 Second-harmonic generation

For d freedoms and N basis states per freedom, the total number of basis states required is of order N^d, so two freedoms are much more of a challenge than one.

Second-harmonic generation or frequency doubling is a standard process of quantum optics which produces squeezed light (see section 10.5). It is a system of two freedoms, which are optical modes or oscillators, with angular frequencies ω_1, ω_2, approximately in 2 to 1 resonance. They interact through a nonlinear coupling in a cavity driven by a coherent classical external field of angular frequency ω_f, which is in near resonance with ω_1. The cavity modes are slightly damped and detuned, with detuning parameters

$$\delta_1 = \omega_1 - \omega_f, \qquad \delta_2 = \omega_2 - 2\omega_f. \tag{6.6}$$

Thus the cavity drives the first oscillator, which then drives the second through the coupling. From the the interaction representation master equation formulation of Drummond and his collaborators [43, 44], the Hamiltonian is given by

$$\left.\begin{aligned} \mathbf{H}/\hbar = {} & \delta_1 \mathbf{a}_1^\dagger \mathbf{a}_1 + \delta_2 \mathbf{a}_2^\dagger \mathbf{a}_2 && \text{(oscillators)} \\ & + i(\mathbf{a}_1^\dagger - \mathbf{a}_1) && \text{(driving force)} \\ & + \tfrac{1}{2}\chi(\mathbf{a}_1^{\dagger 2}\mathbf{a}_2 - \mathbf{a}_1^2\mathbf{a}_2^\dagger) && \text{(coupling),} \end{aligned}\right\} \tag{6.7}$$

where the $\mathbf{a}_j$ are annihilation operators. The damping is represented by the Lindblad operators

$$\mathbf{L}_j = \sqrt{2\kappa_j}\mathbf{a}_j. \tag{6.8}$$

A direct numerical solution of the master equation for this problem is difficult, because in the usual Fock basis the number of basis states needed is equal to the product of the highest significant photon numbers n_j in each mode, and this can become very large. The QSD method with a fixed Fock basis was used by Gisin and Percival [72] and by Goetsch and Graham [78]. Zheng and Savage applied the quantum jump method without a moving basis to the chaotic regime of this system with typical photon numbers (oscillator levels) of 200 in each mode [159]. About 500 basis states per mode were used: roughly 250 000 basis states in all. This was a tour-de-force which required hundreds of hours on a 32-processor supercomputer.

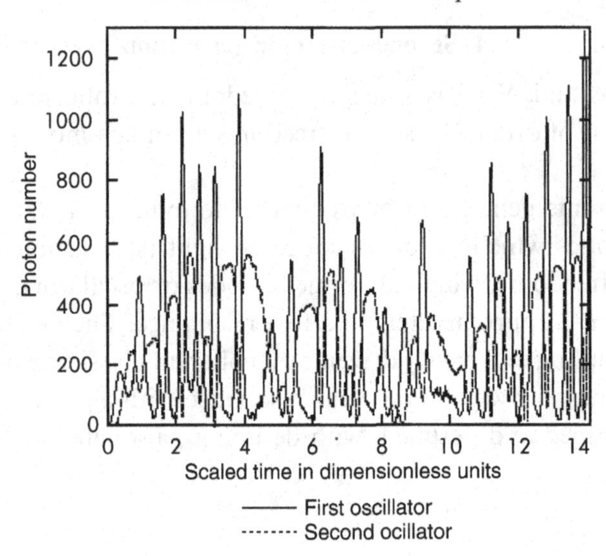

Figure 6.4 The expectation of the photon number versus dimensionless scaled time for a single second-harmonic generation trajectory. Full curve: fundamental mode $\langle \mathbf{n}_1 \rangle$. Broken curve: second harmonic mode $\langle \mathbf{n}_2 \rangle$. The values of the parameters are given in [128].

The same problem was studied using QSD with a moving basis by Schack and his collaborators [127], where more references to earlier treatments may be found. However, the mean photon number was chosen to be about six times larger in each mode, making it even more difficult to solve. The total number of basis states depended on the precision required, but was typically about 200. It took a few hours per trajectory on a PC, and the calculated variation in mean photon number of each oscillator as a function of time is shown in figure 6.4. The big reduction in the number of basis states was due to the use of the moving basis.

In [128], QSD was used on the same problem to compute optical spectra.

6.5 Continuous Stern-Gerlach

In one of the most detailed QSD computations of a laboratory measurement, Steimle and Alber have analysed the continuous Stern-Gerlach effect using QSD with a moving basis [145]. This study is particularly interesting because it makes a connection between the finite measurement times of QSD and the finite measurement times proposed by Dehmelt [27, 28] in his interpretation of experiments on continuous nondestructive measurements of the z-motion and spin of a single electron in a Penning trap. It takes account of environmental effects that are due to the measurement, and others that are not.

In contrast to the original Stern-Gerlach experiment described in section 5.1, the detection process does not absorb the electron, so its dynamics can be monitored continuously, and has been used in experiments to determine fundamental physical constants. The apparatus consists of a Penning trap containing a single electron which undergoes cyclotron, magnetron and z-oscillations of very different frequencies, and an ω_z-shift spectrometer. This has a near-resonant sinusoidal drive which forces the z-motion of the electron into a state of high mean quantum number. This motion is coupled to an external circuit which has a resistor. The dissipation due to the resistor localizes the z-motion into an approximately coherent state whose phase is measured by a phase-sensitive detector. The cyclotron motion is coupled to the electron spin through weak inhomogeneous magnetic fields. The non-dissipative dynamics is represented by a complicated Hamiltonian, which is simplified by using the adiabatic approximation on the hierarchy of frequencies.

The Lindblads represent the effect of the surrounding thermal radiation field on the cyclotron motion, the dissipative effect of the resistor on the electronic z-motion and the continuous measurement of the out-of-phase quadrature component of the current induced in the resistor.

Alber and Steimle's computations used parameters similar to those of the experiments. They showed that it was practicable to vary the effective measurement time, as defined by Dehmelt, over a wide range of values, corresponding to 'complete' and 'incomplete' measurements. They also showed numerically that, in the stationary limit, the ensemble mean of a quantum variance using the QSD wave functions remained essentially unchanged for two different choices of the boundary between system and environment.

This example demonstrates by a tour-de-force calculation how questions of interpretation are related to the practical analysis of real laboratory measurements.

6.6 Noise in quantum computers

Quantum computers are computers that depend on quantum mechanics for their operation [11, 12, 51, 52, 1, 31, 13, 138]. They are very promising [135], but they have not yet been built. The noise produced by interactions with the environment is a major problem. A program based on the QSD library has been used to simulate the effects of noise in quantum computers.

Qubits are introduced in section 3.1. A quantum computer operates on qubits as a classical computer operates on classical bits. An individual qubit is represented in the computer by the state of a single two-state system, such as a spin-half system. The binary numbers b10110 and b11011, representing the decimal numbers d22 and d27, are stored as the product states

$$|d22\rangle = |1\rangle|0\rangle|1\rangle|1\rangle|0\rangle \qquad |d27\rangle = |1\rangle|1\rangle|0\rangle|1\rangle|1\rangle. \tag{6.9}$$

The two numbers are stored together by the *same* five spin systems as the linear combination

$$(1/\sqrt{2})(|d22\rangle + |d27\rangle), \tag{6.10}$$

which is an entangled state of the qubits. More numbers can be stored in more elaborate entangled states up to the 32 dimensions of the Hilbert space of the five spins. Numerical processes on these states are parallel operations on all the numbers. So quantum computers are strongly parallel computers that take full advantage of the dimensionality of the Hilbert space of quantum states. The Hilbert entropy, defined later in section 7.5, measures the information capacity of the binary elements of a quantum computer. Extracting information from a quantum computer is not so simple as extracting it from a classical computer, but it has been proved that if the noise problem can be solved, quantum computers would be far more powerful than classical computers for some problems, like factorization of integers into products of large primes [135]. Quantum error correction methods [136, 143, 144] can be used to reduce the problem, but the correction devices can themselves act as sources of noise.

A multiple spin version of QSD was used by Barenco and his collaborators to support an analytic estimate of the effects of different kinds of noise, which showed that error correction can remain beneficial in the presence of sufficiently weak noise [5]. Their method of error correction for a single qubit over a given lapse time was this. First entangle it, using a coupling Hamiltonian, with four other qubits, then wait for a while with no Hamiltonian, and then disentangle the qubits using another coupling Hamiltonian to produce the wanted output qubit, so that the total time was equal to the lapse time. The noise present during the entire lapse time was represented by Lindblads operating on each of the qubits.

The output state $|\psi'\rangle$ was compared with the input state $|\psi\rangle$ using the 'mismatch' $M = 1 - |\langle\psi|\psi'\rangle|^2$ and the result used to check an analytic estimate, which was found to work well for reasonable values of the parameters.

The difficulty of simulating the quantum computer on a classical machine was itself evidence of the power of quantum computers. In this case it was possible to solve the master equation for 5 qubits, and the comparison with

QSD showed that the latter only had statistical errors, and these were of the order of 1–2%.

As the number n_q of qubits increases, the computer space and time taken by QSD increases approximately linearly with n_q, but for the master equation integration they increase at least quadratically, and soon swamp the space available.

6.7 How to write a QSD program

The QSD program library written by Schack and Brun [130] is available on the World Wide Web. It can be obtained by downloading it from the site

$$\texttt{http://www.ma.rhbnc.ac.uk/applied/QSD.html} \tag{6.11}$$

It is written in the C++ language, but for many purposes only the elements of C++ presented here are needed. The library was used to produce many of the figures of this book. The following description and the program 6.1 show in detail how that was done for figure 4.1a, as an example of how to write QSD programs using the QSD library.

There is no single program. The details of the numerical methods are written in the library, and can be called when needed. For many purposes only a sketchy knowledge of C++ is needed to write programs using the library of main program templates. In C++ new types of variable can be defined, and the usual arithmetic operations like addition and multiplication can be defined for these new types, so programming more closely resembles the usual mathematical notation, and the *use* of the program library has some resemblance to the use of Maple or Mathematica.

The library comes with a number of main programs which can be used as templates to write a wide variety of user programs. They are

```
simple.cc, template.cc, onespin.cc,
spins.cc, moving.cc, sums.cc.                    (6.12)
```

The following shows how to use `onespin.cc` as a template to produce the graph of a Bloch sphere trajectory like that in figure 4.1a.

Figure 4.1a illustrates the evolution of a two-state system, and the natural choice of template is `onespin.cc`, which solves QSD equations for such systems. The user programs are first stored for safety in a separate directory. Then `onespin.cc` is retrieved into the main directory qsd1.3.2 (or later edition) for use as a template. This template is presented as program 6.1.

```
// onespin.cc

// One spin, one trajectory.

#include <math.h>
#include <stdio.h>
#include <iostream.h>
#include <fstream.h>

#include "ACG.h"
#include "Traject.h"
#include "State.h"
#include "Operator.h"
#include "SpinOp.h"
#include "Complex.h"

main()
{
    // Basic operators
  IdentityOperator id;
  NullOperator null;
  SigmaX sx;
  SigmaY sy;
  SigmaZ sz;
  SigmaPlus sp;
  Operator sm = sp.hc();     // sp.hc() is the Hermitian conjugate of sp.

    // The Hamiltonian
  double omega=0.5;
  double epsilon=0.1;
  Operator H = omega*sz + epsilon*sx;

    // The Lindblad operators
  const int nOfLindblads = 1;
  double gamma=0.1;
  Operator L1 = gamma*sm;
  Operator L[nOfLindblads] = {L1};

    // The initial state
  State psi0(2,SPIN);                          // ground state ("spin down")
  psi0 *= sp;                                  // excited state ("spin up")

    // The random number generator
  int seed = 74298;                        // change seed for independent runs
  ACG gen(seed,55);                        // don't change the value 55
  ComplexNormal rand1(&gen);

    // Stepsize and integration time
  double dt=0.01;    // basic time step
  int numdts=10;     // time interval between outputs = numdts*dt
  int numsteps=20;   // total integration time = numsteps*numdts*dt

  double accuracy = 0.000001;

  AdaptiveStep theStepper(psi0,H,nOfLindblads,L,accuracy);
    // deterministic part: adaptive stepsize 4th/5th order Runge Kutta
    // stochastic part: fixed stepsize Euler

// AdaptiveStochStep theStepper(psi0,H,nOfLindblads,L,accuracy);
    // deterministic part: adaptive stepsize 4th/5th order Runge Kutta
    // stochastic part: Euler, same (variable) stepsize as deterministic part

    // Output
  const int nOfOut = 2;

  Operator outlist[nOfOut] = {sz,sx};            // Operators to output

  char *flist[nOfOut] = {"sz.out","sx.out"};     // Output files
    // While the program is running, in addition to the data written in the
    // output files, 7 columns are written to standard output:
    //  the time 't' in column 1;
    //  4 values determined by the array 'pipe' (see below) in columns 2-5;
    //  the effective dimension of Hilbert space in column 6;
    //  the number of adaptive steps taken in column 7.

  int pipe[4] = {1,3,5,7};
    // The 4 numbers in 'pipe' refer to a list formed by the real
    // and imaginary parts of the expectations and variances of the operators
    // in 'outlist'.
    // Example: For outlist={sz,sx} as above, pipe={1,3,5,7} refers to
    // the 1st, 3rd, 5th and 7th entries in the list
    //  { Re(<sz>),Im(<sz>),Re(<sz^2>-<sz>^2),Im(<sz^2>-<sz>^2),
    //      Re(<sx>),Im(<sx>),Re(<sx^2>-<sx>^2),Im(<sx^2>-<sx>^2) },
    // i.e. to the values
    //  Re(<sz>), Re(<sz^2>-<sz>^2), Re(<sx>), Re (<sx^2>-<sx>^2).
```

```
    // Integrate 'nTraj' trajectories, all starting from the same
    //   initial state 'psi0'.
  int nTraj = 1;
  for( int i=0; i<nTraj; i++ ) {
    Trajectory theTraject(psi0,dt,theStepper,&rand1);
    theTraject.plotExp(nOfOut,outlist,flist,pipe,numdts,numsteps);
    cout << endl;
  }
}
```

```
// onespin.cc, edited to produce trajectory like the
// one in figure 4.1a.

// One spin, one trajectory.

#include <math.h>
#include <stdio.h>
#include <iostream.h>
#include <fstream.h>

#include "ACG.h"
#include "Traject.h"
#include "State.h"
#include "Operator.h"
#include "SpinOp.h"
#include "Complex.h"
main()
{
    // Basic operators
  IdentityOperator id;
  NullOperator null;
  SigmaX sx;
  SigmaY sy;
  SigmaZ sz;
  SigmaPlus sp;
  Operator sm = sp.hc();     // sp.hc() is the Hermitian conjugate of sp.

    // The Hamiltonian
  Operator H = null;

    // The Lindblad operators
  const int nOfLindblads = 1;
  double gamma=0.1;
  Operator L1 = gamma*sz;
  Operator L[nOfLindblads] = {L1};

    // The initial state
  double TWOPI=2*3.14159;          // required constant
  Complex I = (0.,1.);             // required constant
  State psi0(2,SPIN);              // ground state ("spin down")
  psi0 = cos(TWOPI/6)*psi0+(I*sin(TWOPI/6))*sp*psi0;
        // In N hemisphere, latitude 30deg
        // sp*psi0 is "spin up".  Scalars multiplied first.

    // The random number generator
  int seed = 74298;                    // change seed for independent runs
  ACG gen(seed,55);                    // don't change the value 55
  ComplexNormal rand1(&gen);

    // Stepsize and integration time
  double dt=0.01;    // basic time step
  int numdts=2;    // time interval between outputs = numdts*dt
  int numsteps=4000;  // total integration time = numsteps*numdts*dt
  double accuracy = 0.000001;

  AdaptiveStep theStepper(psi0,H,nOfLindblads,L,accuracy);

    // Output
  const int nOfOut = 2;
  Operator outlist[nOfOut] = {sy,sz};              // Operators to output

  char *flist[nOfOut] = {"sy.out","sz.out"};     // Output files

  int pipe[4] = {1,5,1,5};

  int nTraj = 1;
  for( int i=0; i<nTraj; i++ ) {
    Trajectory theTraject(psi0,dt,theStepper,&rand1);
    theTraject.plotExp(nOfOut,outlist,flist,pipe,numdts,numsteps);
    cout << endl;
  }
}
```

The final edited form is shown as program 6.2. The output columns of the standard output, sent to a file (using standard output redirection) were used as coordinates for the graph of the trajectory on the Bloch sphere, without the sphere and lines of latitude and longitude.

In figure 4.1a, the Hamiltonian was zero, so it is redefined as

```
H = null;
```

The number of Lindblads is 1, so it is kept as

```
const int nOfLindblads = 1;
```

The value of the coefficient of the Lindblad is 0.135, so the variable gamma is redefined as

```
double gamma = 0.135;
```

To measure the z-spin, the Lindblad is

```
Operator L1 = gamma*sz;
```

When editing the main program, declare every new constant or variable. It must be one of the types already defined in one of the classes, for example a state, a basic operator or an operator. The main programs of the library are a good guide.

Two constants, one real and one complex, are needed to define the initial state:

```
double TWOPI=2*3.14159; // required constant
Complex I = (0.,1.); // required constant.
```

The initial state is closer to the north pole than the south pole, and is given by

```
State psi0(2,SPIN); // ground state (``spin down'')
psi0 = cos(TWOPI/6)*psi0+(I*sin(TWOPI/6))*sp*psi0;
   // In N hemisphere, latitude 30deg
// sp*psi0 is ``spin up''. Scalars multiplied first.
```

The random number generator can be left as it is for one trajectory, but a new seed can be chosen for other trajectories. The basic time interval is sufficiently small to illustrate a trajectory, but for a continuous line output it is advisable to have a shorter time interval between outputs. To illustrate localization, a longer total integration time is advisable. So this section of the program becomes

```
  // Stepsize and integration time
double dt=0.01; // basic time step
int numdts=2;   // time interval between outputs = numdts*dt
int numsteps=4000;
      // total integration time = numsteps*numdts*dt
```
(99)

The output variables, which give the coordinates on the projected Bloch sphere are sz,sx, and so the output becomes

```
// Output
const int nOfOut = 2;
Operator outlist[nOfOut] = {sy,sz} // Operators to output.
```

This gives the standard output, which is used for the graph. For consistency, the output files should also be redefined as

```
char *flist[nOfOut] = ''sy.out'',''sz.out''; // Output files
```

There are two types of output. For each output operator **G** there is an output operator data file with five columns. The first column, labelled 0, is the time t. The rest are $\mathrm{Re}\langle\mathbf{G}\rangle$, $\mathrm{Im}\langle\mathbf{G}\rangle$, $\mathrm{Re}[\langle\mathbf{G}^2\rangle - \langle\mathbf{G}\rangle^2]$, $\mathrm{Im}[\langle\mathbf{G}^2\rangle - \langle\mathbf{G}\rangle^2]$. The columns are treated as if labelled from 0 to 4. This is individual labelling. In all the output operator data files there are $1 + 4*\mathrm{nOfOut}$ independent columns, labelled from 0 to $4*\mathrm{nOfOut}$, with column 0 representing the time, and the remainder being the other columns of the output data files, taken in the order of appearance in the 'outlist' of the program. This is standard output labelling, and is described in the comments which come with the program. The standard output contains a selection from these $1 + 4*\mathrm{nOfOut}$ columns. The first column labelled 0 is the time. The next four are a selection from the output operator data files, which are numbered according to the standard output labelling, as arguments in {, , , } in the '`int pipe[4]`' program line. This part of the standard output can be sent to a file for graphing. Column 6 is '`dim`', the current effective dimension of Hilbert space, that is the number of basis states used at the current time. Column 7 is '`steps`', the current number of adaptive steps taken, which can be used to estimate the number of steps required by the program for a given physical time.

For our standard output, the value in the third column, labelled 2 and representing $\langle \boldsymbol{\sigma}_z \rangle$ is plotted against the value in the second column, labelled 1 and representing $\langle \boldsymbol{\sigma}_y \rangle$.

Further details of the program are given in [130]. More sophisticated changes, like introducing new classes, or programming the classical theory, cannot be made by editing the template main programs, and need a greater understanding of C++.

7

Quantum foundations

This chapter and the next apply the methods of quantum state diffusion to the foundations of quantum mechanics. Problems with the foundations are first introduced and then compared with similar problems in other fields. Quantum state diffusion has developed in parallel with its application to quantum foundations. The relation between them is presented through a selective history of alternative quantum theories that add stochastic terms to the Schrödinger equation. The reasons for introducing these theories, their scope and the constraints that apply to them are discussed.

7.1 Introduction

The foundations of a field of science are not like the foundations of a building, for the first can survive reasonably well without proper foundations, but the second cannot. Nevertheless, the foundations of a field of science can be important, and they are particularly important when the field is being extended into completely new areas. This was true of quantum theory in the 1920s and 1930s and it is true of quantum theory today.

The book by Wheeler and Zurek [154] contains an excellent collection of historical articles on the foundations of quantum mechanics. The quantum theory books by David Bohm [15], by Peres [119] and by Rae [123] use ordinary quantum theory and take particular care with the foundations. Much that is presented in this chapter follows Bell [10].

The quantum dynamics of atoms is very different from the classical dynamics of planets or the dynamics of the classical variables of laboratory apparatus. Seeing this difference in the 1920s, Niels Bohr and his colleagues in Copenhagen and Göttingen divided the universe into the classical and quantum domains, with different physical laws in each. This was an essential feature of Bohr's Copenhagen interpretation of quantum theory.

In Bohr's time, and for most physicists since then, the precise location of the boundary between these domains was of no importance. Atoms, molecules, baryons and leptons are quantum systems, whereas stars, planets, spectroscopes and accelerators are classical. The instruments are around 10^{30} times more massive than the systems they are used to study. The experimenters of the time had no possibility of locating the classical-quantum boundary, so its location was irrelevant in practice.

The Copenhagen interpretation was a glorious and successful compromise, and a powerful theoretical tool for almost all quantum physics since. Classical and quantum systems obey different deterministic dynamics, but there is an essential indeterminacy where they interact. In 1935, those like Einstein and Schrödinger who were dissatisfied with this division of the universe found themselves working at the margins of physics, because their objections made no contact with experiment or observation [47, 131].

Throughout most of the nineteenth century there was doubt about the reality of atoms. According to Ostwald and Mach in 1895, atomic theory was just a convenient model that could be used to answer questions about experiments in chemistry and physics, but those who believed in the reality of atoms were misguided [120]. By the end of the century most scientists had been convinced of their reality, but there were still many who disagreed. Chapter 2 of this book describes how Einstein and Perrin finally resolved the controversy with theory and experiments on Brownian motion. Atoms are real.

Throughout most of the twentieth century there has been doubt about the reality of matter waves, particularly those in more than three dimensions. According to the quantum theory of many textbooks, they are just a convenient model that can be used to answer questions about experiments in physics and chemistry, but those who believe in the reality of these waves are misguided. This view is embedded in the language of those textbooks that tell us about quantum wave functions but not about quantum waves.

But there are now many alternative theories in which all matter consists of quantum waves in three or more dimensions, with very different dynamical behaviour in the classical and quantum limits for large and very small systems. Since matter is real, so are the waves. In some of these theories, there is a diffusion, resembling a Brownian motion, of quantum states. There is today no experimental evidence like Brownian motion for the reality of matter waves, because we still have not been able to experiment at the classical-quantum boundary. However, there is other experimental evidence, as we shall see.

Physical reality was important for both Einstein and Schrödinger. Einstein, Podolsky and Rosen wrote the famous EPR paper on elements of physical reality in 1935 [47]. EPR also argued strongly for the locality of physical influences. John Bell showed that locality and quantum theory were incompatible. Quantum nonlocality predicts distant correlations that are classically impossible, and so could be tested experimentally. Subsequent experiments, notably by Freedman and Clauser in 1972 [53] and by Aspect, Dalibard and Roger in 1982 [3], have provided evidence for quantum nonlocality in the laboratory. Details of these and related experiments are given in [119]. In these situations, physical systems behave nonlocally. So any alternative theory in which quantum systems like atoms and classical systems like measuring apparatus have the same dynamics has to be a nonlocal theory.

Schrödinger was happy with his picture of an electron in a hydrogen atom as a wave in three-dimensional physical space, but not with the two electrons of a helium atom as a wave in a six-dimensional configuration space. He could not accept the reality of the higher-dimensional waves. So in 1935 neither Einstein nor Schrödinger could develop an alternative realistic quantum theory of all matter because of their *other* unshakeable beliefs.

Dirac wrote in 1963 that these problems with quantum foundations were real, but were not then ripe for solution, and so should be left for later [41]. Now modern instruments control individual atoms beyond Bohr's wildest dreams. New experiments are beginning to probe near possible boundaries between the quantum and the classical domains. New theories unite these domains, and some might be tested experimentally. So the problems of quantum foundations may now be ripe.

7.2 Matter waves are real

Today's quantum nonlocality experiments make distant correlations that are classically impossible. Quantum cryptography keeps secrets in nonclassical ways. Section 6.6 explains how quantum computers, if they can be made, will also use higher-dimensional waves. For a long time we have known from Heisenberg's uncertainty principle that we can sometimes do less with a quantum system than expected from classical theory, but now we can sometimes do more. If higher-dimensional quantum waves were not real, how could we use them to do things that can't be done classically?

Higher-dimensional matter waves must be real.

Modern physics demands *universal* theories. Since quantum waves are real and classical systems are real, we need a unified dynamics for both. Without it, there is a gap in our physics, which cannot be filled till we understand the physics at the quantum-classical boundary and which will not be complete without experiments that explore the boundary. Atom traps and interferometers now provide experimenters with remarkable control over individual atoms. The classical-quantum boundary might be in their sights within a few years, so theory is needed to guide the experimenters.

The 'measurement hypothesis' of the usual theory is not enough. We need an alternative theory which describes the dynamics of classical and quantum systems as solutions of the same equations. The distinction between classical dynamics and quantum dynamics must come from limiting properties of the solutions of these equations and not from an evasive interpretation of the theory. Schrödinger's equation and its sophisticated field and string theory extensions have to be modified if they are to represent localized particles that obey classical dynamics. Several alternative quantum theories do this for the Schrödinger equation. For them localization is a dynamical process derived from a modified equation.

We need unified theories of classical and quantum mechanics to guide the experimenters: where should they look for the elusive boundary between classical and quantum systems, and what should they look for?

7.3 Niels Bohr and Charles Darwin

Probability, stochastic processes, and their problems are at least as important in biology as in physics. They are compared in [75]. In the theory of the evolution of species, stochastic processes come in universally as the source of new variation. They also represent the mixing of genes in sexual reproduction, but that does not concern us here.

In Darwin's 1859 theory of evolution, the process of natural selection acts on whatever variability between individuals there is, to produce new varieties and species, but he was not clear on what caused the variability in the first place. The mechanism became clearer with Mendel's 1865 rules of heredity, through what we now call genes, which gradually became accepted after 1900. The genetic theory was successfully incorporated into a general mathematical theory of evolution by natural selection in the 1940s. The new variation is produced by random mutation of genes, which has to be very small over a few generations. What are the genes? It was not until the DNA double helix containing the genes was discovered in 1953, that the physical basis of genetic variation could be understood.

Compare this with quantum theories of measurement.

Although the mathematical laws of probability in the Copenhagen theory are clear, it is no more clear about the mechanism and the physical basis of the indeterminacy of quantum measurements than Darwin was about the mechanism and physical basis of variation in species. Both the biological and physical theories were effective and convincing for the purposes for which they were introduced, and both came to dominate their fields. Both of them were incomplete.

Alternative theories of quantum mechanics with stochastic dynamics are analogous to the mathematical biological theories of the 1940s. In each case the mechanism is clear, but the physical basis of the stochastic fluctuations is not. Just as genetic mutation must be slow on the time scale of generations, so the stochastic dynamics of quantum states is slow on the time scales of the Schrödinger equation. The processes are much more difficult to detect directly than the effects that they produce. As natural selection has led to the development of the beautiful variety of modern species from the simplest beginnings, so the diffusion of quantum states might produce the classical world from the very different quantum world.

The quantum theory equivalent of the DNA double helix would be the experimental detection of quantum fluctuations on an atomic scale, that is, the detection of the quantum-classical boundary. This will be discussed in the next chapter.

7.4 Quantum theory and physical reality

In a letter of 27 May 1926 to Schrödinger on his recently proposed wave equation, H. A. Lorentz pointed out that a wave packet, which when moving with the group velocity should represent a particle, 'can never stay together and remain confined to a small volume in the long run. The slightest dispersion in the medium will pull it apart in the direction of propagation, and even without that dispersion it will always spread more and more in the transverse direction. Because of this unavoidable blurring a wave packet does not seem to be very suitable for representing things that we want to ascribe a rather permanent existence.' This shows the difficulty of using Schrödinger wave packets to represent classical free particles. The dispersion is even stronger when there are particle interactions. This difficulty is overcome by some of the alternative theories.

There are two opposing trends in modern quantum theory.

One is pragmatic, and says that it is enough for a theory to answer well-posed questions about experiments and observations. A theory should be useful, and no more is required. The other says that we should demand more of our theories, and this was the position of Einstein and of Bell, who wanted a complete description of physical reality [47, 10].

John Bell remained to the end of his life a demanding physicist, who said that quantum mechanics was about beables, and not about observables. He gave support to complete theories, in particular the pilot wave theory of de Broglie and Bohm [18, 84] and the theory of Ghirardi, Rimini and Weber [63, 62]. From Bell's viewpoint these are good quantum theories (GQ) because they are complete and because they satisfy four further conditions [10]:

GQ1. The theory can provide a complete and unambiguous description of physical reality.
GQ2. The state and evolution of an individual system are explicitly and unambiguously represented.
GQ3. There is no ambiguous division between the quantum and classical domains.
GQ4. There is no ambiguous division between system and environment.

These conditions are all worth stating, but they are not independent. A theory that satisfies GQ1 must also satisfy the rest, and a theory which satisfies GQ2 must also satisfy GQ3 and GQ4.

Ordinary quantum theory does not satisfy these conditions. It does not satisfy GQ3, and so it does not satisfy GQ1 and GQ2. In the quantum theory of measurement, the system evolves according to quantum mechanics, but the apparatus must obey classical laws if it is to provide a record of the measurement. The division between them is ambiguous. Nevertheless, it is our most successful theory.

Quantum state diffusion for open systems satisfies GQ3, but not GQ4, because the division between system and environment is ambiguous. As it stands, it is not a good quantum theory in the sense of Bell. Thus QSD provides a different representation of a system for different locations of the boundary between system and environment. That is a practical advantage, because the boundary can be adjusted to represent the behaviour of ensembles of open systems by the simplest theory or computation. The predictions are approximate, as for *all* theories when they are applied to specific problems outside cosmology, because it is not possible to include the whole universe, and anything which doesn't is approximate.

But the ambiguity rules out QSD for open systems as a good quantum theory in the sense of Bell. For the example of a quantum measurement, a boundary that is too close to the system does not represent its evolution accurately. A boundary suitable for the computations of chapter 6 gives a good representation of the system, but not necessarily of the measurer. A boundary that includes the important 'classical' variables of the measurer, but not its microscopic variables, provides a good representation of both quantum system and measurer, because the microscopic variables of the apparatus localize the macroscopic variables. But suppose we take the surface of a sphere s light-years away from the quantum system and measurer as the boundary. There is nothing in principle against this, but the quantum state of the system and the classical variables of the environment then remain entangled until the entanglement can be destroyed by interaction with the environment at least $s/2$ years later. Until that time the classical variables remain unlocalized, until then there are unphysical 'Schrödinger cat' states.

There is a wide range of choices of boundary between the extremes that gives a reasonable picture of the system and measurer, but the formulation of a *fundamental* theory should be independent of any such boundary. This is not just a weakness of QSD for open systems, but also of other theories, like those of Zeh [157, 86, 158], and also of Omnès form of consistent histories.

The decoherent or consistent histories theory [80, 101, 60, 42, 61] satisfies GQ3 but not GQ2. Omnès' version does not satisfy GQ4. Nevertheless there are some interesting connections between QSD and the consistent histories approach [40, 81].

If quantum state diffusion is to become a fundamental theory, then the dependence on the boundary must be removed. There must be an *intrinsic* or *spontaneous* diffusion of quantum states. It must be weak in systems which Schrödinger's equation describes well, and strong for classical systems. It must be universal and independent of any environment. Schrödinger's equation must break down somewhere for all physical systems. What contrary evidence do we have that this equation, or its modern extensions to fields and strings, applies universally?

If the Schrödinger equation is to be modified, we need to know how well it has been checked experimentally. How would we check that there are no additional stochastic terms for an arbitrary state of four hydrogen molecules in a 1 cm^3 box, with a maximum energy less than the mean energy at room temperature? We could only do so by detecting the coherence properties of the state. The next section shows that this is not just difficult, it is impossible, because such an arbitrary state cannot be sufficiently well prepared.

7.5 Preparation of quantum states

A system can never be prepared in an exact quantum state. For example, these days the angle of a polarization plate controls the polarization angle of a photon to no better than about 10^{-7} radians. A good measure of the

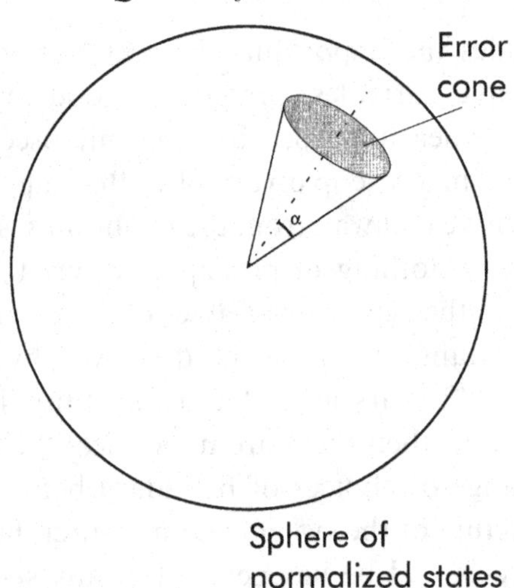

Figure 7.1 Preparation error cone in Hilbert space.

accuracy of preparation is the Hilbert space angle α between the prepared state $|\psi\rangle$ and the target state $|1\rangle$:

$$|\langle\psi|1\rangle|^2 \leq \cos^2\alpha, \tag{7.1}$$

defining a cone around $|1\rangle$ in the Hilbert space, as illustrated in figure 7.1.

The polarization plate is an example of a practical limit to the preparation of a quantum state. There is also a fundamental limit, given by the preparation principle.

Preparation principle. If a 'classical' system C is used to prepare a state of a quantum system Q, then the number of unambiguous preparation states of Q is no greater than the number of distinguishable quantum states of C.

Thus the quantum properties of the 'classical' apparatus put a limit on the precision of preparation of a quantum system. The example of the polarization plate makes this clear. If the angular momentum of the plate is sufficiently well defined, then the precision of the angle of orientation is limited by quantum indeterminacy. This is clearly a fundamental limit on the use of the plate for controlling the angle of polarization of photons. The limit becomes much more severe as the number of possible quantum states increases.

For general systems, if $\alpha \approx \pi/4$ the prepared state is useless for testing the interference properties of Schrödinger's equation, since the uncertainty in the initial state is too great to test them. Suppose the number N of orthogonal quantum states of the system is large. Then the numerical factors in the following can be neglected. Use a logarithmic measure of the number of cells of a partition. This is an entropy in dimensionless units.

Measurement	**Preparation**
Boltzmann entropy S_B	*Hilbert entropy* S_H
Partition phase space into cells (Boltzmann: arbitrary size) (Modern: Planck cells)	Partition sphere of normed states into cells
Number of cells $= N$ (distinct measurement states)	Number of cells $\approx \alpha^{-2N}$ (distinct preparation states)

Putting these relations together we find that the Hilbert entropy for the preparation of a system is approximately equal to the exponential of its Boltzmann entropy:

$$S_B = \ln N, \qquad S_H \approx -2N \ln \alpha \approx 2|\ln \alpha| e^{S_B}. \tag{7.2}$$

We now use the preparation principle, which can be expressed in terms of entropies of the quantum system Q and the classical system C that is used to prepare it as

$$\left.\begin{aligned} &S_H(\mathrm{Q}) \le S_B(\mathrm{C}), \\ &S_B(\mathrm{Q}) \le \ln S_B(\mathrm{C}) \qquad \text{(preparation principle).} \end{aligned}\right\} \tag{7.3}$$

The observable universe is an upper bound on the size of the system C used to prepare Q. Its Boltzmann entropy is $S_B(\text{universe}) \approx 10^{88}$, so if this is used to prepare a quantum system Q, then

$$S_B(\mathrm{Q}) \le 88 \ln 10 \approx 200. \tag{7.4}$$

Figure 7.2 illustrates four hydrogen molecules in a box. The wave number of an H_2 molecule with the mean kinetic energy for a temperature of 300 K is about $3.5 \times 10^{10}\mathrm{m}^{-1}$, giving a quantum number of about $n = 10^8$ for translational motion within a centimetre in one direction. This is illustrated in figure 7.2. So for all three directions there are roughly $n^3 \approx 10^{24}$ states below this energy, and a resultant Boltzmann entropy of translational motion $S_B = 55$ in the dimensionless units used in this section. Consequently four molecules have a greater Boltzmann entropy than 200, and it is impossible to prepare an

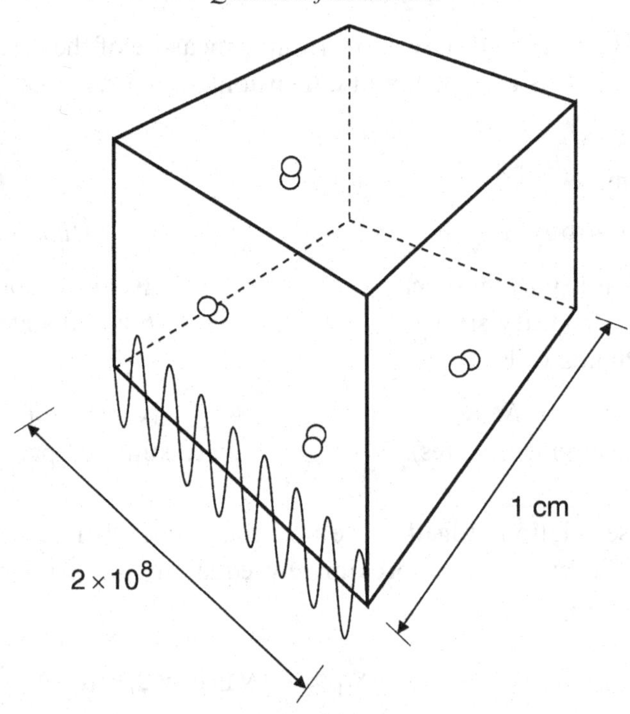

Figure 7.2 States of $\mathbf{H_2}$ molecules in a $1\ \mathrm{cm}^3$ box.

arbitrary state of this system sufficiently well to test its interference properties [110, 129].

Suppose that quantum state diffusion were an intrinsic property of *isolated* quantum systems, and that it was significant for most quantum states of these hydrogen molecules. For most states we have no way to detect the effect of the diffusion on the molecules. We could detect it for some unentangled systems, but there are relatively few of these, and it could be absent for them. The experimental evidence for an unmodified Schrödinger equation is very weak, even for small systems like a few atoms, let alone systems that are normally considered to be classical. There is plenty of room to modify it. The problem is to detect whatever modification there might be.

7.6 Too many alternative theories

Here is a brief account of the early development of the alternative theories that are most closely related to quantum state diffusion. It is not an account of all alternative theories, which would occupy more than the whole book. It does not include theories based on gravity or on spacetime fluctuations, which appear in the next chapter.

In 1966 Bohm and Bub [17] proposed a theory with a space of ordinary state ket vectors $|\psi\rangle$, which represent the state of quantum systems, and satisfy Schrödinger's equation. There is also a dual space of bra vectors which is different from the state vector space and has different dynamics. The components of the bra vectors are to introduce the stochasticity into the physics, but the details of the dynamics are unspecified. Bohm and Bub pointed out that measuring apparatus was just another kind of environment for a quantum system, and should be treated like other environments.

In the same year Nelson put forward his 'stochastic quantum mechanics' in which quantum indeterminacy comes from a sort of Brownian motion of particles in position space on the scale of the usual wave packets [98, 99]. He used the Itô calculus.

In 1976 Pearle [103] introduced nonlinear terms into the Schrödinger equation for an individual quantum system, in which apparently random variations are produced by small changes in the phase factor of the state. In a measurement there is a resulting diffusion towards the eigenstates. In 1979 [104] he wrote an article with the title 'Towards explaining why events occur', which very clearly stated the need for reality of the waves of quantum dynamics, and introduced an explicit stochastic equation for these waves.

In 1984, Gisin [68, 67] considered the measurement of a two-state system, represented by the projector $\mathbf{P}$, and contrasted the von Neumann postulate for the density operator

$$\rho \rightarrow \mathbf{P}\rho\mathbf{P} + (\mathbf{I} - \mathbf{P})\rho(\mathbf{I} - \mathbf{P}), \tag{7.5}$$

which is a linear relation, with the Lüders postulate for the state vector

$$\left.\begin{array}{ll} |\psi\rangle \rightarrow |\mathbf{P}\psi\rangle/||\mathbf{P}\psi|| & \text{with probability } \langle\psi|\mathbf{P}|\psi\rangle \\ |\psi\rangle \rightarrow |(\mathbf{I} - \mathbf{P})\psi\rangle/||(\mathbf{I} - \mathbf{P})\psi|| & \text{with probability } \langle\psi|\mathbf{I} - \mathbf{P}|\psi\rangle, \end{array}\right\} \tag{7.6}$$

which is nonlinear. He showed that Lüders transition follows from the long time limit of the solution of a nonlinear stochastic Itô equation of QSD form, but without the complex fluctuations, demonstrating localization to an eigenstate. He derived the von Neumann postulate as a limit of the linear master equation for the density operator obtained from the fluctuating pure states. These letters demonstrated the importance of a consistent evolution of the master equation, which he analysed in more detail in 1989 [69], as described in section 7.7. In 1986 Diósi [32] showed how to unravel a dissipative master equation using quantum jumps.

Also in 1986, Ghirardi, Rimini and Weber [63] produced the first complete theory, with a universal stochastic localization of position. They demonstrated how the localization could be undetectable for small quantum systems, yet dominant for large classical ones. The rates were specified by two undetermined parameters. However, the theory was based on equations for density operators, not pure states, and on jumps. Bell [10] presented the theory in terms of stochastic equations for the states, and it was later reformulated in terms of diffusing quantum states by Diósi [34], Gisin [69], Pearle [107] and Ghirardi, Pearle and Rimini [64], who gave it the name of continuous spontaneous localization.

In 1988 Diósi extended Gisin's state diffusion theory to include position localization. He showed that a free particle localizes to a Gaussian wave packet that behaves like a Brownian particle [34, 35]. He also generalized the theory for wide open systems in terms of the probability distribution $\mathbf{Pr}(\psi)$ of section 3.2, with an alternative representation in terms of orthogonal quantum jumps [36].

These theories all suppose some intrinsic, spontaneous diffusion or jumps, and satisfy Bell's conditions. They all have real fluctuations in the sense of section 4.6, not the complex unitary invariant fluctuations of QSD introduced in [72] and used in chapter 4 to obtain the QSD equation from the master equation.

The theories which follow the early work of Ghirardi, Rimini and Weber are based on Lindblads which depend on the position. However, there is strong numerical and other evidence from QSD that phase space localization to wave packets is a property of all systems with both a Hamiltonian and a sufficiently strong Hermitian Lindblad or Lindblads, except for very simple or special cases. The Lindblad produces localization in only one direction in the $2d$-dimensional phase space, but the Hamiltonian then mixes this direction with the others, so that the quantum wave localizes in all directions in the phase space. This process is very clearly demonstrated for the example in section 4.3 of a mechanical oscillator whose Lindblad is proportional to the momentum. It localizes in both momentum and position, producing an oscillating wave packet with near minimum indeterminacy product. A set of mechanical oscillators like atoms in a solid, with momentum localization, will localize in phase space, and therefore in position. Collisions between particles have the same effect, so atoms in gases with a variety of Hermitian Lindblads will also localize in phase space. Phase space localization includes position localization, so, for real systems with Hamiltonians that include interaction, position localization does not need Lindblads that depend on position.

The energy localization of the next chapter *is* very special. For nonrelativistic fluctuations, it does not localize in position.

Intrinsic state diffusion could occur for entangled systems as simple as four hydrogen molecules at 300 K without violating any possible experiments, let alone actual experiments. Conseqently there is an *enormous* range of Lindblads and parameters for intrinsic QSD that are consistent with experience and satisfy Bell's conditions.

The problem is not to find an alternative theory, but to choose among this variety of possible theories. How then can we guide experimenters to possible violations of Schrödinger's equation so that they can detect the quantum-classical boundary? The next section describes an important constraint on possible theories, but the variety remains. The next chapter suggests an answer.

7.7 Gisin condition

Shimony [134] uses 'peaceful coexistence' to describe the resolution of the difficult problems of compatibility of the nonlocality of quantum mechanics and special relativity. In the next chapter we show that not all of them have been resolved.

A condition on compatibility of realistic nonlinear evolution equations and special relativity was put forward by Gisin [68] as part of a lively exchange of comments with Pearle [105], who subsequently gave a detailed treatment of the special case of matrix diagonalization [106]. This section mainly follows Gisin's 1989 paper [69], whose methods he [70] and Polchinski [122] used in a critique of Weinberg's nonlinear quantum theory [151, 152].

In chapters 3 and 4 we accepted the experimental evidence that the evolution of an ensemble of quantum systems depends on the density operator ρ and not on its unravelling into pure states. We now show that this fundamental property is a consequence of special relativity, that signals cannot be transmitted faster than the velocity of light, and also an additional assumption about preparation of entangled systems. It follows from these conditions, and from the linearity of the evolution of density operators, that any deterministic evolution of the state vectors must also be linear.

Suppose the density operator $\rho(0)$ of a system A at time $t = 0$ has two unravellings into pure states $|i\rangle$ with projectors $\mathbf{P}_i(0)$ and into states $|j\rangle$ with projectors $\mathbf{P}_j(0)$, where there are m different values of i and $n \leq m$ different values of j

$$\rho(0) = \sum_i \mathbf{Pr}(i)\mathbf{P}_i(0) = \sum_j \mathbf{Pr}(j)\mathbf{P}_j(0) \qquad (\mathbf{Pr}(i) \neq 0,\ \mathbf{Pr}(j) \neq 0). \quad (7.7)$$

The states with projectors $\mathbf{P}_i(0)$ need not be orthogonal, and the same is true of the $\mathbf{P}_j(0)$. None of the probabilities $\mathbf{Pr}(i)$, $\mathbf{Pr}(j)$ is zero.

Since the sets of states $|i\rangle$ and $|j\rangle$ make the same density operator they must span the same state space, and since none of the probabilities is zero, we can find real amplitudes c_i, c_j and a matrix V_{ij} such that

$$\left.\begin{aligned} c_i^2 = \mathbf{Pr}(i), \qquad & c_j^2 = \mathbf{Pr}(j), \\ c_i|i\rangle = \sum_j V_{ij} c_j |j\rangle. & \end{aligned}\right\} \quad (7.8)$$

Substituting into the equations (7.7) for the density operator then gives

$$\sum_{ijj'} c_j V_{ij} |j\rangle\langle j'| V^*_{j'i} c_j{}' = \sum_j c_j^2 |j\rangle\langle j|, \qquad \text{so} \qquad \sum_i V^*_{j'i} V_{ij} = \delta_{j'j}. \quad (7.9)$$

For $m = n$ the matrix V_{ij} is unitary, and in general it is a rectangular generalization of a unitary matrix.

Suppose that there is a distant physical system B with a set of m orthonormal states $|u_i\rangle$ and another set of n states

$$|v_j\rangle = \sum_i V^*_{ji} |u_i\rangle, \quad (7.10)$$

whose orthonormality follows from the generalized unitary property (7.9) of V_{ij}. Then the entangled state

$$|\chi\rangle = \sum_i c_i |i\rangle|u_i\rangle = \sum_j c_j |j\rangle|v_j\rangle \quad (7.11)$$

of the combined system AB has a reduced density operator

$$\rho_{\mathrm{A}}(0) = \mathrm{Tr}_{\mathrm{B}} |\chi\rangle\langle\chi|. \quad (7.12)$$

If $|u_i\rangle$ and $|v_j\rangle$ are nondegenerate eigenstates of two operators **Y** and **N** for the system B, then a **Y** measurement of B unravels the system A in one way and an **N** measurement unravels it the other way.

Now suppose, contrary to what we want to prove, that the evolution of pure states is deterministic and nonlinear, and that the evolution of the two unravellings can lead to different density operators at a later time t. Denote the deterministic evolution by the possibly nonlinear (super)operator $\mathbf{T}(t)$ on the projectors $\mathbf{P}$ such that

$$\mathbf{T}(t)\mathbf{P}(0) = \mathbf{P}(t). \tag{7.13}$$

Suppose that the two ensembles corresponding to the two unravellings evolve after time t into ensembles with *different* density operators

$$\rho_1(t) = \sum_i \mathbf{T}(t)\mathbf{P}_i(0) \neq \rho_2(t) = \sum_j \mathbf{T}(t)\mathbf{P}_j(0). \tag{7.14}$$

Suppose the entangled system $|\chi\rangle$ can be prepared with A and B further apart than ct where c is the velocity of light. Suppose an experimenter Barbara at B wants to transmit a binary message to an experimenter Arthur at A faster than the velocity of light: 'yes, I will', or 'no, I won't'. Then she has to measure $\mathbf{Y}$ for yes or $\mathbf{N}$ for no at time 0. The result is that the system A goes into one or other of the unravelled ensembles. Arthur patiently waits for a time t, and then measures the expectation of an operator $\mathbf{G}$ for which

$$\mathrm{Tr}\mathbf{G}\rho_1(t) \neq \mathrm{Tr}\mathbf{G}\rho_2(t). \tag{7.15}$$

The probability of a given g depends on the unravelling, and so it depends on Barbara's message. This changes the probability of 'yes' and of 'no', but doesn't give a definite answer. The probability of an unfortunate misunderstanding can be reduced to an arbitrarily small value by carrying out many copies of the experiment in parallel.

Physics prevents even the most passionate messages being sent faster than the velocity of light, and so *if* the state χ can be prepared, it follows that the evolution of the density operators cannot depend on the unravelling. As a direct result of this it follows that the operator $\mathbf{T}(t)$ of (7.13) applies also to the density operators, whose evolution is linear. Consequently the evolution of the projectors must be linear, and that means that the evolution of the states is also linear.

We conclude with Gisin that if signals cannot be sent faster than the velocity of light, and if entangled states of the type $|\chi\rangle$ can be prepared, then deterministic nonlinear evolution of state vectors is impossible.

We have seen in section 7.5 that there are strong limitations on the preparation of entangled states for all but the simplest systems, so we can only conclude that the deterministic nonlinear evolution contradicts special relativity for these. But it would be difficult to construct a theory with strictly linear evolution for simple systems and nonlinear evolution for more complicated systems, so, whilst deterministic nonlinear evolution of state vectors is not entirely ruled out, it is unlikely.

Gisin and Rigo have taken this argument further, by showing how to add precise noise terms to the deterministic nonlinear evolution equation which give solutions closest to the solutions of the master equation [76].

8

Primary state diffusion – PSD

Primary state diffusion, or PSD, is obtained twice, from different principles. First comes a nonrelativistic formulation that follows from the need to derive the long-time behaviour of a physical system from its short-time behaviour [112, 113]. The second formulation relates PSD to spacetime fluctuations, and follows a long tradition that relates alternative quantum theories to gravity [114, 115]. The second formulation has no free parameters, but it is not unique, because of different possible geometries of fluctuations in spacetime. Experimental tests are proposed, which might also serve as tests of quantum gravity theories. The chapter concludes with the implications for quantum measurement and a discussion of the many remaining unsolved problems. Some of these issues are discussed in [116, 115].

8.1 First approach – Schrödinger from diffusion

The first form of primary state diffusion is derived in this section as the simplest formulation of a nonrelativistic theory in which quantum state diffusion is a fundamental property of all physical systems [112]. This theory leads to energy localization, a good approximation for systems that remain sufficiently localized in space, but the nonrelativistic theory is incomplete because there is no space localization of extended systems.

In science, the evolution of a system over a long time period can be obtained by combining the effects over shorter time periods. If diffusion and drift are both present, then for short periods the diffusion dominates the drift. If they are sufficiently short, the drift is negligible compared with the diffusion. This leads to a reversal of the usual roles of diffusion and Schrödinger dynamics, whereby the long-time Schrödinger dynamics is derived from the short-term diffusion dynamics. The diffusion is fundamental and universal, and we derive the drift from the diffusion, not the other way round.

Primary state diffusion means that the quantum state diffusion is primary and the drift, including ordinary quantum mechanics, is secondary. In the first version, the principles of primary state diffusion (PSD) are:

PSD1. *State vector*. The state of a system at any time t is represented by a normalized state vector $|\psi(t)\rangle$. This applies to classical systems like the solar system, pointers, data records and Brownian particles as well as to quantum systems like electrons and photons.

PSD2. *Operator*. The dynamics of the system is determined by a system diffusion operator $\mathbf{K}$, which is linear.

PSD3. *State diffusion*. The change of state obeys an *elementary* QSD equation, with Lindblad $\mathbf{K}$. There is no Schrödinger term.

$$|\mathrm{d}\psi\rangle = \langle\mathbf{K}\rangle^*\mathbf{K} - \tfrac{1}{2}\mathbf{K}^\dagger\mathbf{K} - \tfrac{1}{2}\langle\mathbf{K}\rangle^*\langle\mathbf{K}\rangle)|\psi\rangle\mathrm{d}t + \mathbf{K}_\Delta|\psi\rangle\mathrm{d}\xi. \tag{8.1}$$

The operator $\mathbf{K}$ is not Hermitian and $\mathbf{K}_\Delta = \mathbf{K} - \langle\mathbf{K}\rangle$, as in equation (3.22). The quantum fluctuations of PSD3 are independent of spatial coordinates in the nonrelativistic theory, but are space-dependent in any relativistic theory. The operator $\mathbf{K}$ determines the Schrödinger dynamics along with the state diffusion.

The choice of $\mathbf{K}$ is severely constrained by the following conditions.

C1. *Schrödinger evolution*. Over longer periods, when state diffusion is negligible, the Schrödinger evolution is given by the anti-Hermitian part of the drift operator in the QSD equation, so the quantum state diffusion equation must be of the form

$$|\mathrm{d}\psi\rangle = -(i/\hbar)\mathbf{H}|\psi\rangle\mathrm{d}t + \mathbf{R}|\psi\rangle\mathrm{d}t + \mathbf{K}_\Delta|\psi\rangle d\xi, \tag{8.2}$$

where the first term represents the regular Schrödinger evolution and the second term is the *residual* drift.

C2. *Limited disruption*. The diffusion is not so fast that it disrupts the regular Schrödinger evolution to a degree that violates experimental or observational evidence.

C3. *Fast diffusion*. The rate of diffusion is sufficiently fast to localize classical dynamical variables.

The last condition is based on the assumption that there *are* dynamical variables that behave classically, and that this classical behaviour does not normally depend on acts of observation or measurement.

For systems described by classical variables, the state vectors must be so localized in configuration and phase space that they cannot be distinguished from classical points which move according to the classical equations of motion. In particular the motion of the centre of mass of a planet, stone or Brownian particle is classical.

Notice that conditions C2 and C3 are opposing restrictions on PSD theory, one ensuring that the state diffusion is not too fast, and the other that it is not too slow.

By equating corresponding terms in the state diffusion equation of PSD3 and equation (8.2) of C1, we find that for all $|\psi\rangle$,

$$\begin{aligned}\langle\psi|\mathbf{K}_\mathrm{R}|\psi\rangle\mathbf{K}_\mathrm{I} - \langle\psi|\mathbf{K}_\mathrm{I}|\psi\rangle\mathbf{K}_\mathrm{R} &= c(\psi)\mathbf{I} - \mathbf{H}/\hbar \\ &= c(\psi)\mathbf{I} - \langle\psi|\mathbf{I}|\psi\rangle\mathbf{H}/\hbar,\end{aligned} \tag{8.3}$$

$$\mathbf{R}(\psi) = -\tfrac{1}{2}\,\mathbf{K}_\Delta^\dagger(\psi)\mathbf{K}_\Delta(\psi), \tag{8.4}$$

where $\mathbf{I}$ is the unit operator, $c(\psi)$ is a real constant corresponding to an arbitrary phase factor, and $\mathbf{K}_\mathrm{R}, \mathbf{K}_\mathrm{I}$ are Hermitian. So $c(\psi)$ must have the form

$$c(\psi) = \langle\psi|\mathbf{C}|\psi\rangle \tag{8.5}$$

for some operator $\mathbf{C}$. Since the equation is valid for all $|\psi\rangle$, it can be written as an equation for outer products of operators, as

$$\mathbf{C}\otimes\mathbf{I} - \mathbf{I}\otimes\mathbf{H}/\hbar = \mathbf{K}_\mathrm{R}\otimes\mathbf{K}_\mathrm{I} - \mathbf{K}_\mathrm{I}\otimes\mathbf{K}_\mathrm{R}. \tag{8.6}$$

Since the right side is antisymmetric, so is the left side, and

$$(\mathbf{H}/\hbar)\otimes\mathbf{I} - \mathbf{I}\otimes(\mathbf{H}/\hbar) = \mathbf{K}_\mathrm{R}\otimes\mathbf{K}_\mathrm{I} - \mathbf{K}_\mathrm{I}\otimes\mathbf{K}_\mathrm{R}. \tag{8.7}$$

$\mathbf{K}$ must be a linear combination of the Hamiltonian and the unit operator $\mathbf{I}$, and since the dynamics is independent of a phase factor of unit modulus multiplying $\mathbf{K}$, the coefficient a_1 of $\mathbf{H}$ can be taken as positive.

Substitution shows that the real part of the coefficient of $\mathbf{I}$ has no effect on the dynamics, so

$$\mathbf{K} = a_1\mathbf{H} + ia_2\mathbf{I}, \tag{8.8}$$

where a_1 is positive and a_2 is real, so that by (8.7)

$$\mathbf{K} = a_1\mathbf{H} + \frac{i}{\hbar a_1}\mathbf{I}. \tag{8.9}$$

When this is substituted into the original QSD equation it leads to a *transformed* QSD equation

$$\begin{aligned} |\mathrm{d}\psi\rangle = (-i/\hbar)\mathbf{H}_\Delta dt - a_1^2\mathbf{H}_\Delta^2|\psi\rangle \mathrm{d}t \\ + a_1\mathbf{H}_\Delta|\psi\rangle \mathrm{d}\xi \qquad \text{(transformed).} \end{aligned} \tag{8.10}$$

Note that as $a_1 \to 0$, the diffusion becomes negligible and the transformed QSD is the Schrödinger equation, apart from a nonphysical phase factor. This equation of nonrelativistic PSD has the same form as the state diffusion equation for the measurement of energy. So the principles PSD1–PSD3 and the condition C1 lead to a universal intrinsic energy localization. This was first proposed by Bedford and Wang [7, 8], tentatively suggested as an intrinsic state diffusion process by Gisin [69], and appeared as an approximation to a density operator decoherence theory in the work of Milburn [95] and Moya-Cessa et al. [97].

In the extreme case of a single Hamiltonian for the whole universe, the fluctuation $\mathrm{d}\xi$ acts simultaneously over all space at a given time. This satisfies Galilean relativity, but is an obvious violation of special relativity, for which there is no absolute simultaneity. Nevertheless the nonrelativistic theory gives a good picture of what happens to a system of nonzero rest mass in a limited region of spacetime.

In PSD, as in Brownian motion, there is a time constant τ_0 that marks the boundary between the time intervals for which diffusion dominates and those for which the drift dominates. This is given by

$$\tau_0 = a_1^2\hbar^2, \qquad a_1 = \tau_0^{1/2}/\hbar. \tag{8.11}$$

The time constant τ_0 is universal for PSD, and more useful than a_1.

8.2 Decoherence

Ramsey devised a method of putting an atom into a coherent linear combination of two different energy states with energy difference $\Delta E = \hbar\omega$. Since PSD produces energy localization, these combinations have a fundamental characteristic lifetime, after which the diffusion reduces the standard deviation in energy to less than ΔE, so that the system tends to go into one or other of the energy states.

Using the general theory of localization, this gives a 'primary decoherence rate' Γ_P,

$$\Gamma_P = \omega^2 \tau_0, \tag{8.12}$$

whose inverse is a primary coherence decay time. The decay time increases as the square of the energy difference. If a Ramsey experiment does *not* show decoherence, this puts an upper bound on τ_0. This leads to a theory of decoherence rates.

Primary state diffusion is not the only possible reason for decoherence. It can be dissipated earlier by decay, like the radiative decay of an atom, or the radioactive decay of a nucleus, or by interaction with matter in the environment, as in line broadening, after which the coherence is no longer observable. Normal environments are sufficiently big for this secondary decoherence by dissipation to mask the primary decoherence, and the rate for this will be denoted Γ_S, the secondary coherence decay rate.

Experiments are not normally designed to measure decoherence rates, but nevertheless many do so, because unwanted decoherence interferes with the purpose of the experiments. Because all relevant experiments to date have been found consistent with ordinary quantum theory,

$$\Gamma_S > \Gamma_P \tag{8.13}$$

and

$$\tau_0 < \frac{\Gamma_S}{\omega^2} = \tau_\chi, \tag{8.14}$$

where τ_χ is a characteristic time for the secondary decoherence, which is different for each experiment, and puts a bound on the universal time constant τ_0. Specific bounds are given in [112].

8.3 Feynman's lectures on gravitation

In his lectures on gravitation (1961–2) [50], Richard Feynman said

> The extreme weakness of quantum gravitational effects now poses some philosophical problems; maybe nature is trying to tell us something here, maybe we should not try to quantize gravity. ... It is still possible that quantum theory does not absolutely guarantee that gravity *has* to be quantized. ... In this spirit I would like to suggest that it is possible that quantum mechanics fails at large distances and for large objects. Now, mind you, I do not say that I think that quantum mechanics *does* fail at large distances, I only say that it is not inconsistent with what we know. If this failure is connected with gravity, we might speculatively expect this to happen for masses such that $GM^2/\hbar c = 1$, or M near 10^{-5} g.

This is the Planck mass, which is given below in (8.15).

He continues later:

> If there was some mechanism by which the phase evolution had a little bit of smearing in it, so it was not absolutely precise, then our amplitudes would become probabilities for very complex objects. But surely, if the phases did have this built in smearing, there might be some consequences to be associated with this smearing. If one such consequence were to be the existence of gravitation itself, then there would be no quantum theory of gravitation, which would be a terrifying idea for the rest of these lectures.

This theme was taken up by Károlyházy in 1966 [88, 87], who connected stochastic reduction of the wave function with gravity through an imprecision in the spacetime structure, but the imprecision comes from the presence of large masses, and the resultant decoherence is very weak.

There the matter rested for many years.

In 1993 Diósi and Lukács [39, 38] showed that the Fourier expansion of the small scale fluctuations in Károlyházy's theory leads to unacceptably large fluctuations in energy density. The same objection applies to spacetime PSD, the theory introduced in this section. However, the objection is based on the Fourier relation between spacetime and energy-momentum. Section 4.2 shows that in QSD these Fourier relations are incompatible with special relativity, and so must break down for PSD and related theories. Thus the objection cannot be sustained.

Nevertheless this result of Diósi and Lukács shows how important it is to find a replacement for these Fourier relations.

Hawking [82] made a connection between entropy loss in black holes and quantum decoherence, which was used as the basis of Penrose's picture of an alternative quantum theory [108]. A wave function that is sufficiently spread out in space, and gets coupled to a larger system which produces significant

Weyl curvature, represents a significant rise in the gravitational entropy, and then reduction takes place.

During 1987–9, Diósi introduced explicit stochastic gravitational theories of localization [33, 37], emphasizing that both gravity and quantum mechanics must be changed because neither is valid on all scales. He based his theory on a classical fluctuating Newtonian gravitational field. Ghirardi, Grassi and Rimini [66] showed that this theory requires a further fundamental length, which they provided in a modified theory.

Ellis, Mohanty and Nanopoulos modified the ordinary Schrödinger evolution, with non-unitarity coming from wormhole interactions [48].

Again, as in the previous chapter, there is a wide range of possible theories, in some of which there are free parameters. Some of them have parameters whose values can be so extreme that there is no accessible critical experiment that is guaranteed to distinguish them from ordinary quantum theory, which is frustrating for those who might want to test them experimentally. For example, any theory requiring a test of the interference properties of particles with mass close to the Planck mass is inaccessible to experiment [116].

8.4 Second approach – spacetime PSD

Planck introduced his fundamental time, length and mass shortly before his first paper on quantum theory. In their modern form, using $\hbar$ for the Planck constant, they are

$$\left.\begin{aligned} T_{\text{Planck}} &= (\hbar G/c^5)^{\frac{1}{2}} \approx 5.39 \times 10^{-44}\,\text{s}, \\ L_{\text{Planck}} &= (\hbar G/c^3)^{\frac{1}{2}} \approx 1.62 \times 10^{-35}\,\text{m}, \\ M_{\text{Planck}} &= (\hbar c/G)^{\frac{1}{2}} \approx 2.18 \times 10^{-8}\,\text{kg}, \end{aligned}\right\} \tag{8.15}$$

where G is Newton's gravitational constant.

In spacetime PSD, primary state diffusion is derived from different foundations [114]. Although the Schrödinger equation is primary and the fluctuations are secondary, the name is retained because the theory is the same. The quantum fluctuations are obtained from stochastic fluctuations of spacetime on and below the scale of the Planck length and time. This fixes the scale of the time constant τ_0. On these scales the transformation from a flat Minkowski universe to our universe resembles the transformation from the time variable to the space variable of Brownian motion. This transformation is not differentiable and requires the Itô differential calculus to replace ordinary differential calculus. There is a close parallel with the application of the Itô calculus to Brownian motion in chapter 2.

Although special and general relativity are used for the properties of spacetime, the spacetime PSD equation itself is nonrelativistic. It is proposed as the nonrelativistic limit of some unknown relativistic theory, discussed in the concluding section of this chapter. The principal experimental test of PSD is in atom interferometry, for which the atoms move at speeds of metres per second, and the usual relativistic corrections are utterly negligible.

In the original PSD equation given above, the fluctuations $d\xi$ are coupled to the quantum system through the Hamiltonian; that is, relativistically, they are coupled to matter through the mass. But the gravitational field is also coupled to matter through the mass. This suggests that quantum fluctuations and gravity might be different manifestations of the same field. In general relativity, gravity comes from the large-scale structure of spacetime, so we should expect the quantum fluctuations to come from the small-scale structure of spacetime.

This results in a semiclassical theory in which matter is treated quantally, and spacetime is represented classically. The probabilistic quantum properties of matter follow from the stochastic spacetime fluctuations, which drive the quantum state through the resultant diffusion term which is added to the Schrödinger equation. With this hypothesis, the only important constants to appear are the velocity of light, the gravitational constant G and the Planck constant $\hbar$ for quantum mechanics. In effect, there are no free parameters.

According to the equivalence principle of general relativity, around every spacetime point of our universe there is a sufficiently small region with a smooth transformation relating it to a flat Minkowski universe. If spacetime has quantum fluctuations on the Planck scale, and they are independent along timelike paths, then the transformation from a flat Minkowski spacetime to our spacetime resembles Brownian motion on larger scales, and the transformation is not differentiable. So there are no locally inertial Lorentz frames in our universe, and no small region around any point with a smooth transformation relating it to a Minkowski universe. The equivalence principle does not hold. As we go down in scale towards the Planck domain, spacetime become less flat, not more so. Spacetime PSD is based on a stochastic differential structure and a fluctuating spacetime metric whose quantum fluctuations are expressed in terms of nondifferentiable relations between classical spacetimes. The usual quantum fluctuations of QSD are consequences of these spacetime fluctuations.

Spacetime PSD theory is therefore made of two parts: a Minkowski universe and its Lorentz frames, and the properties of our universe that are obtained by local nondifferentiable spacetime transformations between the Minkowski universe and ours.

In a Lorentz frame of the Minkowski universe with time $\bar{t}$, nonrelativistic quantum systems are represented by states that satisfy the Schrödinger equation,

$$\hbar|\mathrm{d}\psi\rangle = -i\mathbf{H}\mathrm{d}\bar{t}|\psi\rangle \tag{8.16}$$

to a good approximation. Quantized fields satisfy the flat spacetime laws of quantum field theory. There is no classical dynamics because there is no quantum state diffusion and no localization. In PSD, as in other theories that satisfy Bell's conditions for a good quantum theory, there is only one world, which includes both quantum and classical dynamics, but for PSD the existence of the latter depends on the localization in phase space produced by quantum state diffusion.

As for Brownian motion, the physics depends on the scale. On sub-Planck scales, the transformation from flat Minkowski spacetime to the spacetime of our universe is dominated by the fluctuation term. On far super-Planck scales it is dominated by the smooth differentiable transformations of general relativity, which for simplicity are replaced by the identity transformation. On intermediate scales a proper time interval Δs for a timelike segment of spacetime is represented by

$$\Delta s = \Delta\bar{s} + \tau_1^{\frac{1}{2}}\Delta\xi(x), \tag{8.17a}$$

where τ_1 is a universal time constant and the fluctuation depends on spacetime points, not just on the time. This is a conformal transformation, in which the Minkowski metric is multiplied by a scalar factor, as considered by Sánchez-Gómez [126]. The spacetime transformations and the new metric are both complex. An imaginary component is needed to produce classical dynamics from quantum dynamics, but it makes no other discernible change on macro scales.

For nonrelativistic systems, the differential time-time component of the fluctuating transformation is then given to a good approximation by

$$\mathrm{d}\bar{t} \approx \mathrm{d}t + \tau_1^{\frac{1}{2}}\mathrm{d}\xi. \tag{8.17b}$$

The simplest assumption is that τ_1 is the Planck time T_{Planck}, but a small numerical factor C of order unity, like 2π, cannot be excluded, so

$$\tau_1 = CT_{\text{Planck}}. \tag{8.18}$$

Apart from the constant C, spacetime PSD is a rigid theory with no free parameters. Near the Planck scales, the transformation between the Minkowski universe and our universe is complex. It is this which produces the linear fluctuation term of the QSD equation. The nonlinear terms of the QSD equation are needed to preserve the norm and produce the localization.

These are the principles of spacetime PSD:

STP1. The state vector $|\psi\rangle$ satisfies a nonlinear QSD equation, where the right side is an operator on $|\psi\rangle$.

STP2. On far sub-Planck scales the operator is given by the transformation from the flat Minkowski universe, except for an additional scalar times the unit operator, which is needed to preserve normalization.

STP3. On far super-Planck scales the anti-Hermitian part of the operator is given by the transformation from the Minkowski universe. This is the Schrödinger operator.

STP4. The norm of $|\psi\rangle$ is preserved.

These principles lead to the following consequences.

Using the Schrödinger equation in Minkowski space and STP1-3, the state vector satisfies an equation of the form

$$\hbar|\mathrm{d}\psi\rangle = (-i\mathbf{H}\mathrm{d}t + \mathbf{R}(\psi)\mathrm{d}t + \tau_1^{1/2}\mathbf{H}\mathrm{d}\xi - s(\psi)\mathbf{I}\mathrm{d}\xi)|\psi\rangle, \tag{8.19}$$

where $\mathbf{R}(\psi)$ is an Hermitian operator.

It follows from STP4 that

$$\begin{aligned} 0 = \mathrm{d}\langle\psi|\psi\rangle = {} & 2\mathrm{Re}\langle\psi| - i\mathbf{H} + \mathbf{R}(\psi)|\psi\rangle\mathrm{d}t \\ & + 2\mathrm{Re}(\langle\psi|\tau_1^{1/2}\mathbf{H} - s\mathbf{I}|\psi\rangle\mathrm{d}\xi) \\ & + \langle\psi|(\tau_1^{1/2}\mathbf{H} - s\mathbf{I})^\dagger(\tau_1^{1/2}\mathbf{H} - s\mathbf{I})|\psi\rangle\mathrm{d}t \end{aligned} \tag{8.20}$$

and since $\mathrm{d}\xi$ can vary, its coefficient must be zero, so $s = \tau_1^{1/2}\langle\psi|\mathbf{H}|\psi\rangle$ and

$$\mathbf{R}(\psi) = -\tfrac{1}{2}\tau_1\mathbf{H}_\Delta^2. \tag{8.21}$$

So the state vector in the uniformly fluctuating universe satisfies the primary state diffusion equation

$$\hbar|\mathrm{d}\psi\rangle = (-i\mathbf{H}_\Delta\mathrm{d}t - \tfrac{1}{2}\tau_0\mathbf{H}_\Delta^2\mathrm{d}t + \tau_0^{\frac{1}{2}}\mathbf{H}_\Delta\mathrm{d}\xi)|\psi\rangle, \tag{8.22}$$

with

$$\tau_0 = \tau_1 = CT_{\text{Planck}}, \tag{8.23}$$

as required.

8.5 Geometry of the fluctuations

For Brownian motion the diffusion constant was enough to specify the motion uniquely for scales large compared with atomic scales. Unfortunately the specification of the time constant τ_0 is not enough to specify the fluctuating spacetime transformation of PSD uniquely. The problem is that we do not know the geometry of the fluctuations, that is the form that the fluctuations take as a function of space and time. There are many possible Lorentz-invariant geometries. Any of the main theories of quantum gravity, such as superstring theory, or the noncommutative geometry of Connes, might be used to obtain the statistical properties of the fluctuations in spacetime. But they have not yet been derived. Here we choose three possible simple geometries that appear to be reasonable [117].

G1. Independent fluctuations in spacetime Planck cells of four-volume $T^4_{\text{Planck}}c^3$, not to be confused with the more usual Planck cells in phase space.

G2. Fluctuations propagating with the velocity of light, with commutative geometry.

G3. As G2, but with *non*-commutative geometry.

We choose the simplest example of each. The main purpose is to find out if the fluctuations might be seen in the laboratory.

8.6 Matter interferometers

Einstein's theories of spacetime and relativity depend on the comparison of proper times for two clocks that follow different paths between two spacetime points. The clocks start together, move apart, and then come together again. This is the kind of experiment that is needed to detect spacetime fluctuations, but it is not nearly accurate enough.

Nowadays we can do much better. An atom is a quantum clock, with a very high frequency, proportional to its mass. (This way of using an atom as a clock is quite different from its use for frequency standards.) In an atom interferometer the *same* atom follows the two different paths, 'here' and 'there', with a state given by the coherent linear combination

$$\frac{1}{\sqrt{2}}|\text{clock here, vacuum there}\rangle + \frac{1}{\sqrt{2}}|\text{vacuum here, clock there}\rangle. \quad (8.24)$$

This increases the accuracy so much that atom interferometry seems to offer the best chances of detecting spacetime fluctuations in the laboratory. The reason is that the energy difference in the decoherence rate Γ given by (8.12) is the rest mass energy mc^2 of the atom, so

$$\Gamma = (mc^2/\hbar)^2\tau_0, \qquad \Gamma_{u,\mathrm{Planck}} = 1.08 \times 10^5 \mathrm{s}^{-1}, \quad (8.25)$$

where $\Gamma_{u,\mathrm{Planck}}$, the decay rate for an atomic mass unit u when $\tau_0 = T_{\mathrm{Planck}}$, is experimentally accessible. The phase change of each wave packet is proportional to the proper time along its path, resulting in constructive or destructive interference where they recombine. A phase change of π in either path interchanges constructive and destructive interference. So the interference pattern tells us something about the time difference between the paths.

In the absence of fluctuations, the phase change over a proper time interval T is $\eta(T) = A\Omega_u T$, where Ω_u is the frequency for the atomic mass unit and A is the atomic number of the atom. The intrinsic spacetime fluctuations add an additional fluctuating phase, whose value is obtained by multiplying $\eta(T)$ by the fluctuation factor given in section 2.1:

$$\delta\eta(T) = (\tau_0/T)^{1/2}\eta(T) = A\Omega_u(\tau_0 T)^{1/2}. \quad (8.26)$$

If the fluctuations were to produce phase shifts of about 1 radian, they would reduce the contrast of the interference pattern. In practice the effect can be detected with a phase shift ten or a hundred times smaller.

In matter interferometry it is difficult to avoid interactions with the environment which also reduce the contrast. The challenge is to detect the spacetime fluctuations unambiguously. From the formulae for $\delta\eta(T)$ it follows that the best experiments are those with relatively large atomic number A and large atomic drift time T.

For example, in the interferometer built by Kasevich and Chu [89], the centres of the wave packets for individual sodium atoms ($A = 23$) were kept about 6 mm apart for 0.2 s. Using the formula, $\eta(T) \approx 7 \times 10^{24}$ radians. A detectable phase change in $\eta(T)$ of 0.1 radians represents a precision of about 1 part in 10^{26} for the proper times in this experiment. This extraordinary precision has already been achieved, and offers the possibility of detecting fluctuations in spacetime through the reduction in the contrast of the interference pattern.

When the phase shift $\eta(T)$ is scaled by the fluctuation factor $(\tau_0/T)^{1/2}$, with τ_0 equal to the Planck time, the size of the fluctuations is approximately 3000 radians. This is much greater than the detectable 0.1 radians, so independent Planck scale spacetime fluctuations in the two paths would easily destroy the interference pattern.

The experimental bound on the time constant τ_0 for the three geometries above has been obtained [117] for an atom of atomic number A, a drift time T and a spacetime area $c\mathcal{A}$ between the paths (thus $\mathcal{A}$ is in units of time^2). When the relatively small kinetic energy terms were neglected, the results for the three geometries were as follows.

G1. Independent:

$$\tau_0 < \frac{10^{-49}\,\mathrm{s}^2}{2A^2T} \quad (\approx 1.8 \times 10^{-51}\,\mathrm{s}). \tag{8.27}$$

G2. Propagating, commuting:

$$\text{none.} \tag{8.28}$$

G3. Propagating, noncommuting (spin-half):

$$\tau_0 < \frac{10^{-49}\mathrm{s}^2}{A^2\mathcal{A}^{1/2}} \quad (\approx 1.8 \times 10^{-46}\mathrm{s}). \tag{8.29}$$

The approximate numerical values are for the experiment of Kasevich and Chu.

But the experimenters did observe interference. There was a reduction in contrast, but this could be explained as the result of interactions with the environment, giving bounds on the time constants for the first and third cases which are actually less than the Planck time. The main weakness of this argument is that the model of the interferometer is fairly crude. No account has been taken of the finite size of the atomic wave packets in the experiments. Nevertheless it demonstrates that the small numerical value of the Planck time does not alone prevent experimental access to Planck-scale phenomena. The small value of the Planck time is not an impassable barrier to the precision of modern matter interferometers.

This experiment (and several others) give evidence against some types of spacetime fluctuation. It would be much better to see positive evidence of spacetime fluctuations through suppression of interference, which could be tried by increasing the atomic number A, the atomic drift time T and possibly the separation of the wave packets. It would also be necessary to remove the effects of environmental interactions as far as possible.

8.7 Conclusions

Given a geometry of spacetime fluctuations, primary state diffusion defines the physics of the classical-quantum boundary unambiguously. If we assume that the time constant τ_0 for the fluctuations is within a few orders of magnitude of the Planck time, then, for the relatively simple geometries chosen, either there is no localization, or it is so fast that it should already have been observed as a reduction in the contrast of the interference pattern in matter interferometry experiments. There is still no unambiguous evidence for this reduction of contrast. It is too early to conclude that the experiments rule out the theory, because the spacetime fluctuations used were not derived from any theory of quantum gravity, and the model of the interferometer was fairly crude. Other geometries that are derived from quantum gravity theories might give a weaker localization that is compatible with experiment.

So far, the predicted rate of localization easily satisfies the other bound, since it is certainly fast enough to localize macroscopic systems and Brownian particles.

Special relativistic invariance poses severe problems for realistic localization theories. Experiments on Bell inequalities show a form of nonlocality that is reproduced in nonrelativistic primary state diffusion through the nonlinear terms in the equations. To my knowledge no one has succeeded in finding a realistic Lorentz invariant theory of this type. Conversely, despite the obvious difficulties with nonlocality, no one appears to have shown conclusively that such a theory is impossible. The linear unravelling of section 4.5 is compatible with special relativity [65], but it provides no explicit and unambiguous representation of the evolution of an individual system, so it does not satisfy Bell's condition GQ3 for a good quantum theory. We know that in any good Lorentz invariant quantum theory based on QSD the usual Fourier relations between conjugate variables like position and momentum must break down, but there is still no such theory.

It has been known for a long time that quantum gravity implies spacetime fluctuations. Fluctuations in the early universe may have produced the observed inhomogeneities in the distribution of galaxies and in the universal background radiation today. But there is still no final theory which unites gravity and the other forces of nature, despite the enormous theoretical effort that has been put into producing one. There is not even a universally accepted theory of quantum gravity.

So, although numerical evidence suggests that spacetime fluctuations could become accessible to experiment using matter interferometers, there is no guarantee. It depends on the nature of the fluctuations, and this depends on quantum gravity, which is not fully understood. The preliminary estimates presented here indicate that some types of spacetime fluctuations are ruled out by experiment. This suggests that feasible experiments might distinguish between different fundamental theories, such as various versions of superstring theory, or noncommuting geometries of spacetime. This would be a significant advance in a field which needs more experimental checks, if it can be done. It may rule out some theories, but we don't know with certainty what effects, if any, that they predict, so we can't yet tell.

Despite the arguments presented here, it is not certain that the problem of the classical-quantum boundary in realistic quantum theories is related to spacetime fluctuations, but so far they appear to be the only way of limiting the enormous variety of localization processes discussed in section 7.6 to a manageable choice. There is supporting evidence in this chapter that two completely distinct lines of development lead to the same theory. The first, in which the quantum state diffusion is primary, and Schrödinger evolution is derived from it, has one free parameter. The parameter is determined by the second, which depends on spacetime fluctuations.

This suggests that the classical-quantum boundary and spacetime fluctuations are different aspects of the same thing, that the detection of one would also be the detection of the other, that the problems of quantum foundations and quantum gravity are linked, and the solution of one could bring with it the solution of the other. If this is so, then *both* theories would have to change. We have discussed at length how quantum theory has to change if the evolution of individual systems is to be represented explicitly. We have presented evidence that the basic equations for quantum amplitudes are nonlinear, nonlocal, and preserve the norm. Because they are nonlinear they are non-unitary. Unitarity is a cornerstone of theories like superstring theory, even when many other familiar properties of everyday theoretical physics are abandoned. This is an indication of how much these theories would have to change.

9

Classical dynamics of quantum localization

In the laboratory, on scales large compared with Planck's constant, localization is well represented by a nonlinear stochastic dynamics of quantum densities in classical phase space. This is quantum density diffusion dynamics. As in Hamiltonian dynamics, which is a more familiar large-scale limit of Schrödinger dynamics, the wave properties of quantum systems are not represented. The action takes place in phase space as a generalization of the Liouville dynamics of phase space distributions. This chapter introduces the dynamics of quantum densities descriptively through the conditions that it must satisfy. Its derivation as a classical limit of QSD is given in the next chapter. Localization properties of the dynamics are derived. The phase space density diffusion theory presented here corresponds to the quantum theory of localization by Hermitian Lindblads in QSD given in chapter 5, but it is simpler, and the theory is taken further.

9.1 Introduction

Quantum state diffusion provides a dynamics for the localization or reduction of a quantum mechanical wave packet during a measurement or similar physical process. The essence of this localization dynamics is captured in the classical theory presented here, which applies to the common situation where the localization takes place over macroscopic distances and the wave properties of the system are no longer relevant. It provides a picture of the localization process in classical phase space. In this chapter the theory is developed from classical Hamiltonian dynamics in phase space and the known properties of localization. It is derived as part of the semiclassical limit of QSD in the next chapter. Localization theorems are proved, and the theory is used to show that absorption by a screen leads to quantum jumps.

In a Stern-Gerlach experiment, the detection process produces localization over macroscopic distances. This is typical of quantum measurements, whether they are laboratory measurements, or the more general type of measurement defined in section 3.5, and suggests a classical localization theory.

Hamiltonian dynamics can be derived as a limit of Schrödinger dynamics, but it can also be treated as an independent theory in its own right. Here we show that the essence of localization of quantum states in phase space can be treated similarly. It can be captured in a classical picture, which is formulated as the localization of densities in classical phase space. Physical processes are easier to visualize in phase space than in Hilbert space. The properties of density localization are obtained from the known properties of localization dynamics in QSD. From these properties we obtain the basic density diffusion equations, which the next chapter shows to be the classical limit of QSD.

In many situations, and particularly for measurement processes, this classical model of localization describes the essential physics. It has also proved useful both for practical applications of QSD, and for an understanding of the role of localization in measurement theory. In the numerical applications of chapter 6, localization is used to make large reductions in the number of basis states. This reduction relies on the phase space picture, which is much easier to analyse by using density diffusion theory than using QSD itself. In measurement theory, the density diffusion provides a simpler picture of localization by measurement, in the common situations where the wave properties of the quantum states are no longer relevant. This is illustrated by the simplified example of absorption by a screen presented in section 9.9.

Quantum state diffusion is obviously very different from classical Hamiltonian dynamics. There are two features in particular that distinguish them. One is that in QSD matter has wave properties, such as interference and diffraction, and the other is the property of localization. The Schrödinger equation captures the first, but not the second. The density localization theory introduced here captures the second, but not the first. The relations between the theories are illustrated in figure 9.1.

For simplicity the density diffusion theory of this chapter is restricted to nonrelativistic systems whose interactions with the environment are represented by Hermitian Lindblad operators in QSD. These include interactions with measurement apparatus and similar interactions.

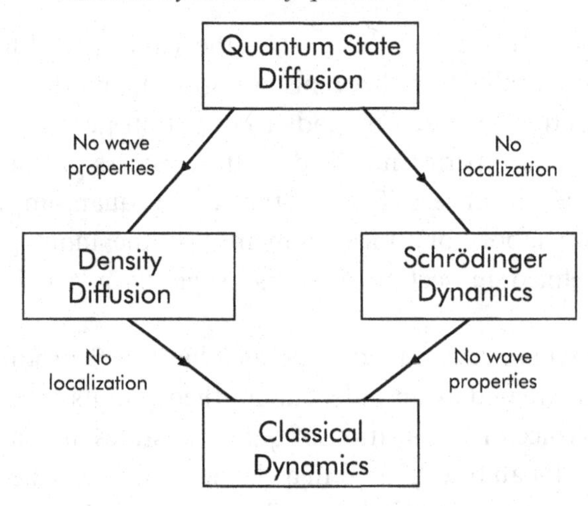

Figure 9.1 Relations between theories.

Sections 9.2 and 9.3 introduce the basic properties of the phase space densities of individual quantum systems, whilst section 9.4 deals with ensembles. A density diffusion equation for simple systems is obtained in section 9.5, and it is generalized in section 9.6. The next two sections provide measures of localization: entropy in section 9.7 and variance in section 9.8. Section 9.9 applies the theory to localization of a particle in a beam that strikes a solid surface, providing a paradigm for quantum measurement, and showing how quantum jumps can be derived from this classical version of quantum state diffusion. Section 9.10 discusses density diffusion as a physical process and as a new type of classical dynamics.

This chapter is based on [118].

9.2 Classical systems and quantum densities

A quantum particle, unlike a classical particle, can be in more than one place at a time. It is nonlocal.

The state of a single classical system with d freedoms is represented by a point (q, p) in phase space, where q and p are d-dimensional vectors representing the coordinate and conjugate momentum. A single quantum system without its wave properties, like interference and diffraction, is represented here by a *quantum density* $D(q, p)$ in classical phase space, which is never negative. Because it does not represent the wave properties of a quantum system, the quantum density should not be used where the wave properties dominate. The quantum density represents the nonlocality.

There is an analogy in ordinary ray optics, where on large scales the light intensity is given by tracing the rays of light, but this does not give the wave properties, like interference and diffraction. The density D is analogous to the light intensity, which is a coarse-grained electromagnetic energy density.

Ordinary ray optics does not need a theory of localization, but for individual photons it does. When a partially absorbent lens focuses a very weak broad-band photon source of finite size onto an absorbent screen, the lens and screen both localize the photons. A photon then has a quantum density D which is a function of its position and momentum. The energy density in phase space is proportional to the mean over the quantum densities for the individual photons, and the density diffusion theory of this chapter applies.

The quantum density D is not the same as the Wigner distribution W, which retains the wave properties, and shows structure down to the scale of Planck cells of volume $(2\pi\hbar)^d$. The next chapter treats the relation between their equations of motion. The quantum density D has none of the characteristic wave properties like interference and diffraction, and shows structure only on larger scales. The phase space dynamics of phase space densities represents quantum localization on scales much larger than the Planck constant $\hbar$, which is where the classical dynamics of phase space points is used to represent Schrödinger dynamics. Indeed, in the limit when localization of the phase space densities has taken place, density diffusion dynamics is the same as classical dynamics.

The dynamics of a classical system is represented by a phase space trajectory $(q(t), p(t))$. For Hamiltonian dynamics with Hamiltonian $H(q, p, t)$, the trajectory is determined by Hamilton's equation

$$(\dot{q}, \dot{p}) = (\partial H/\partial p, -\partial H/\partial q), \tag{9.1}$$

where the partial derivatives represent gradients in momentum and coordinate space. The probability of a state of an ensemble in classical statistical mechanics is represented by an ensemble of points in phase space whose probability distribution is $\rho(q, p)$. When the system is isolated, so that there is no interaction with the environment, the value of $\rho(q(t), p(t))$ remains constant as the state moves along the Hamiltonian trajectory. We say that the phase point is carried by the Hamiltonian flow in the phase space. The time dependence of the distribution $\rho(q, p)$ as a function of fixed points (q, p) is given by the linear Liouville equation

$$\frac{\mathrm{d}\rho(q,p)}{\mathrm{d}t} = \sum_i \left(\frac{\partial\rho}{\partial q_i}\frac{\partial H}{\partial p_i} - \frac{\partial\rho}{\partial p_i}\frac{\partial H}{\partial q_i} \right) = \{\rho, H\}, \tag{9.2}$$

where $\{\rho, H\}$ is a Poisson bracket.

Quantum systems are different. Matter interferometers show that a single electron, neutron or atom can be in two or more places at once. A quantum system cannot therefore be represented by a point in phase space. However, a single quantum system in a pure state without its wave properties can be represented by a non-negative normalized density $D(q,p)$ such that

$$\int_\Omega D(q,p,t) = 1, \tag{9.3}$$

where $\int_\Omega$ represents an integral over all phase space. Its value at (q,p) is proportional to the conditional probability that the system will be found by a measurement in a region of phase space around (q,p). The region must be large on the scale of Planck's constant, but can be small on macro scales.

For quantum particles we have to work with both densities D and distributions ρ in phase space, and keep the distinction between them. The density is a large-scale smoothed approximation to the phase space distribution of a pure quantum state. Dynamical variables have density expectations, variances and covariances, which are approximations to quantum expectations, variances and covariances. These have the same form as the corresponding quantities for the distributions, but must not be confused with them. A density and a density expectation are quantum properties of the system, even though their representation is classical.

The density D is a conditional probability distribution, like those of section 3.5. It is not a probability distribution like ρ, although when there is no localization it has the same dynamics. Quantum wave properties, like interference and diffraction, are not represented, and so a quantum interference pattern is averaged over many neighbouring maxima and minima. Localization properties *are* represented.

The loss of the wave properties provides a classical picture of the physics of quantum localization and also makes the analysis simpler than for QSD, because many drift terms disappear. A particle in a beam can be represented by a density D that is spread throughout some volume of the beam, which is often of macroscopic dimensions. It is also spread in momentum. A free electron in a solid at room temperature is represented by a density D distributed within the solid. An electron in a beam which strikes the surface of the solid may be represented by a density that is partly in the beam, and partly in the solid, an important case for us, which is treated in section 9.9.

The point density

$$D(q, p) = \delta(q - q_0, p - p_0) = \delta(q - q_0)\,\delta(p - p_0) \tag{9.4}$$

represents a quantum state confined to a region around the phase space point (q_0, p_0), which is negligibly small on macro scales. It represents a wave packet that is confined in phase space on a macro scale, although it may still extend over regions that are large on the scale of powers of a Planck cell.

Without localization, the density $D(q, p, t)$ satisfies the same classical Liouville equation as the distribution $\rho(q, p, t)$ of phase space points. Localization introduces additional terms. Point densities are already localized, so the localization has no effect on them, and they follow Hamiltonian trajectories in phase space. The classical picture of quantum localization has the same strengths and weaknesses as ordinary classical dynamics. It no longer works on scales determined by Planck's constant, and does not represent the wave properties of material particles, such as interference and diffraction.

Since localization is a stochastic process, the density $D(q, p, t)$ satisfies a stochastic equation in time. The probability distribution $\mathbf{Pr}(D)$ of densities D represents an ensemble of densities. This probability is a function of points in the space of densities, which is a large space, like the Hilbert space of quantum mechanics. Fortunately it is not used very much.

9.3 Quantum expectations and other properties of densities

The density D represents the quantum state of a single system of an ensemble. The integrated value of D over a region of phase space is a *quantum weight*, representing a conditional probability, not a probability, although it is related to a probability through the stochastic dynamics of localization. More detail is given in section 3.5. Dynamical variables have quantum expectations, given by weighted averages over the whole phase space Ω.

The simplest example is the norm $N(D)$ of D, which is the expectation of unity,

$$N(D) = \int_\Omega D(q, p) \cdot 1. \tag{9.5}$$

The norm must always be equal to 1.

Quantum expectations, quantum variances and quantum covariances are defined in terms of these densities, just like the corresponding means, variances and covariances of classical probability distributions. The quantum density *expectation* of a dynamical variable G is

$$\langle G \rangle = \langle G \rangle_D = \int_\Omega D(q,p) G(q,p). \tag{9.6}$$

The square of the variation of G around the expectation $\langle G \rangle$ is the quantum density *variance*

$$\sigma^2(G) = \sigma(G,G) = \langle (G - \langle G \rangle)^2 \rangle = \langle G^2 \rangle - \langle G \rangle^2. \tag{9.7}$$

The ensemble localization $\Lambda(G)$ of G is defined to be the inverse of the mean over the ensemble of the quantum density variance of G:

$$\Lambda(G) = 1/\mathrm{M}\,\sigma^2(G). \tag{9.8}$$

The quantum density *covariance* of two dynamical variables G and F is

$$\sigma(G,F) = \langle (G - \langle G \rangle)(F - \langle F \rangle) \rangle = \langle GF \rangle - \langle G \rangle \langle F \rangle. \tag{9.9}$$

For ordinary classical probability distributions, the covariance is a measure of the correlation of G and F, but for these densities it is a measure of the quantum *entanglement* of G and F. Because of localization, the properties of entanglement are essentially different from the properties of correlation. In fact it is primarily entanglement that makes quantum systems behave so strangely. Our use of 'entanglement' here is a generalization of its normal use in quantum mechanics, as in section 3.3.

The density dispersion entropy S_D of a density D is minus the expectation of the logarithm of the density:

$$S_D = -\langle \ln D \rangle = -\int_\Omega D(q,p) \ln D(q,p). \tag{9.10}$$

It is a useful logarithmic measure of the dispersion of the density D in phase space. The more the density is dispersed throughout the space, the larger the entropy, the faster the density localizes in phase space, the faster the entropy decreases. Unlike the quantum dispersion entropy of section 5.6, the density dispersion entropy does not depend on any choice of partition of the state space into channels.

9.4 Probability distributions and means

Like phase space points for open classical systems and the state vectors of quantum systems in QSD, phase space densities for open quantum systems obey statistical laws. A quantum system that interacts with a measurer is open, and it is this type of interaction which concerns us.

Given a single initial density D, the future densities cannot be predicted exactly, but they have a probability distribution $\mathbf{Pr}(D)$, so there is a stochastic dynamics of quantum densities. In density dynamics, the distribution $\rho(q,p)$ is defined in terms of the probabilities $\mathbf{Pr}(D)$. If there is only one density $D(q,p)$ in the ensemble, then $\rho(q,p) = D(q,p)$. When there are more, ρ is an ensemble mean M over densities:

$$\rho(q,p) = \int \mathrm{d}(D)\,\mathbf{Pr}(D) D(q,p) = \mathrm{M}\,D(q,p), \tag{9.11}$$

where the integral is over the whole D-space. If the probability distribution ρ is derived from the probability distribution of densities $\mathbf{Pr}(D)$, then the ensemble mean of the expectation for a dynamical variable G is

$$\mathrm{M}\,G = \mathrm{M}\,\langle G\rangle = \int \mathrm{d}(D)\,\mathbf{Pr}(D)\langle G\rangle = \int \mathrm{d}(D) \int_\Omega D(q,p)G(q,p). \tag{9.12}$$

This will be called the ensemble mean, or even just ‘the mean’ of G. Unlike the expectation, it is a property of ensembles. In the special case when all the D are point densities, ρ is the usual phase space distribution of phase points for the particles, and the ensemble mean is the usual ensemble mean of classical statistical mechanics. We will show that the dynamics of localization tends to make systems approach this special case.

The mean (9.12) is linear in D. However we will be particularly interested in the means of variances, like $\mathrm{M}\,\sigma^2(G)$, because they can be used to measure the localization for an ensemble of densities. These are quadratic in D. Quadratics in D also appear in the localization dynamics of densities, which is therefore nonlinear, unlike the Liouville equation (9.2).

For quantum particles we have to work with both quantum densities and distributions in phase space, and keep the distinction between them. The quantum density expectations, variances and covariances must not be confused with the corresponding quantities for the distributions. The quantum expectation and variance, like the density itself, are quantum properties of the system, even though their representation is classical.

9.5 Elementary density diffusion

At first, suppose that the system is simple, so that the Hamiltonian is negligible and the diffusion depends on a single dynamical variable $L = cF$, corresponding to an Hermitian Lindblad operator in QSD. This represents the localization of F as a result of density diffusion, where c is a nonzero constant that determines the rate of localization. The measurement of F is an example of such an interaction. The fundamental quantum density diffusion equations which localize the densities need to satisfy a number of reasonable conditions. The simplest possible linear equation satisfies all the conditions except the normalization condition, which can easily be restored by introducing a quadratic normalization term. A number of important results follow.

The first two conditions apply to the density D itself. They are

Co1. The norm $N(D)$ of D is conserved.
Co2. The non-negativity of D is preserved.

The rest are conditions on the mean and localization of dynamical variables.

Co3. The ensemble mean of any G remains constant.
Co4. The dynamical variable L localizes.

The only condition that may be surprising is Co3, which is not true for QSD. However, equation (5.16) shows that the variation of $\mathrm{M}\langle \mathbf{G}\rangle$ is zero if the commutators are zero, which they are in the classical limit.

To obtain the density diffusion equations, first assume for simplicity an initial ensemble at time t with only one density $D(t)$. At later times the ensemble will have many densities. Because of Co2, $D(t)$ is real for all times and the fluctuations must be real. Real fluctuations are denoted $\mathrm{d}w = \mathrm{d}w(t)$, where

$$\mathrm{M}\,\mathrm{d}w(t) = 0, \qquad \mathrm{M}\,(\mathrm{d}w(t))^2 = \mathrm{d}t \tag{9.13}$$

and the fluctuations at different times are statistically independent, the Markov condition of QSD.

Let M be the mean over these fluctuations, as usual. The simplest non-trivial diffusion equation for a density D which includes the Lindblad dynamical variable L and the fluctuation $\mathrm{d}w$ is

$$\mathrm{d}D = LD\mathrm{d}w \qquad \text{(trial equation).} \tag{9.14}$$

This equation is linear in D, but it cannot be right, because the norm is not conserved:

$$\mathrm{d}N(D) = \langle L \rangle \mathrm{d}w \qquad \text{(trial).} \tag{9.15}$$

This can be rectified by subtracting the expectation of $\langle L \rangle$ from L, giving

$$\mathrm{d}D = (L - \langle L \rangle)D\mathrm{d}w \qquad \text{(density diffusion).} \tag{9.16}$$

This is the simplest density diffusion equation. The right side of the evolution equation is quadratic in D, and, because it is nonlinear, the density $D(q,p)$ cannot be a probability distribution. The expectation $\langle L \rangle$ depends on the value of $D(q,p)$ in all regions of phase space for which it is not zero. So the change in D depends on its value in all these regions. This usually means that the evolution is nonlocal, as quantum localization is known to be. So this density diffusion equation is nonlinear in D and nonlocal.

We now demonstrate that the solution of (9.16) satisfies all the conditions Co1–Co4.

For the normalization condition Co1,

$$\begin{aligned} \mathrm{d}N &= \int_\Omega \mathrm{d}D = \int_\Omega (L - \langle L \rangle)D\mathrm{d}w \\ &= (\langle L \rangle - \langle L \rangle)\mathrm{d}w = 0. \end{aligned} \tag{9.17}$$

For Co2, if the value of $D(q,p)$ at any point (q,p) is zero, then it remains unchanged, and so by continuity D cannot pass the value zero.

For Co3, the change in the expectation of an arbitrary dynamical variable G is

$$\begin{aligned} \mathrm{d}\langle \mathrm{G} \rangle &= \int_\Omega G\mathrm{d}D \\ &= \int_\Omega G(L - \langle L \rangle)D\mathrm{d}w = \langle G(L - \langle L \rangle)\rangle \mathrm{d}w \\ &= \sigma(G, L)\mathrm{d}w, \end{aligned} \tag{9.18}$$

so, for arbitrary G,

$$\mathrm{M}\, \mathrm{d}\langle G\rangle = 0, \tag{9.19}$$

which confirms Co3.

For Co4, we need particular cases of (9.18) and (9.19) for $G = L$ and $G = L^2$. These are

$$\mathrm{d}\langle L\rangle = \sigma^2(L)\mathrm{d}w, \qquad \mathrm{M}\, \mathrm{d}\langle L^2\rangle = 0. \tag{9.20}$$

Using the Itô differential of a product (2.14), the ensemble mean of the change in the variance of L is

$$\begin{aligned} \mathrm{M}\, \mathrm{d}\sigma^2(\mathrm{L}) &= \mathrm{M}\, \mathrm{d}\langle L^2\rangle - \mathrm{M}\, \mathrm{d}\langle L\rangle^2 \\ &= -\mathrm{M}\, (\mathrm{d}\langle L\rangle)^2 = -\mathrm{M}\, (\sigma^2(L))^2 \mathrm{d}t. \end{aligned} \tag{9.21}$$

From the inequality (5.12) it follows that

$$\mathrm{d}\Lambda/\mathrm{d}t \geq 1, \qquad \Lambda(t) \geq \Lambda(0) + t. \tag{9.22}$$

The mean of the quantum variance of L decreases, unless it is already zero. The factor of 2 between the rates for density diffusion and QSD can be traced to the different normalization conventions for real and complex fluctuations.

We will now show that any dynamical variable G is localized unless the covariance $\sigma(G, L)$ is zero, and so L deserves the name of *localizer*. From equations (9.18) and (9.19), we have for arbitrary G that

$$\mathrm{M}\, \mathrm{d}\sigma^2(G) = -\mathrm{M}\, (\mathrm{d}\langle G\rangle)^2 = -(\sigma(G, L))^2 \mathrm{d}t, \tag{9.23}$$

from which the result follows.

An important consequence is that a localizer proportional to a coordinate q localizes a conjugate momentum p unless their covariance is zero. Only when the localization approaches the Heisenberg limit does the quantum commutator become significant, so that the indeterminacy principle is not violated. Before that time the covariance term dominates.

9.6 Generalization

A general density diffusion equation has a Hamiltonian H and many localizers L_j with their corresponding fluctuations $\mathrm{d}w_j$:

$$\mathrm{d}D = \{D, H\}\mathrm{d}t + \sum_j (L_j - \langle L_j \rangle) D \mathrm{d}w_j, \tag{9.24}$$

where $\{D, H\}$ is the usual Poisson bracket for the Hamiltonian evolution, and the fluctuations $\mathrm{d}w_j$ are independent and normalized to $\mathrm{d}t$, so that

$$\mathrm{M}\,\mathrm{d}w_j = 0, \qquad \mathrm{M}\,\mathrm{d}w_j \mathrm{d}w_k = \delta_{jk}\mathrm{d}t, \tag{9.25}$$

as in section 2.9. These orthonormal fluctuations $\mathrm{d}w_j$ are not unique. Any orthogonal transformation gives an equivalent set $\mathrm{d}w'_k$ of orthonormal fluctuations $\mathrm{d}w_j = \sum_j O_{jk}\mathrm{d}w'_k$. The localizers L_j may also be considered as a vector, in the space of dynamical variables. The orthogonal transformation $L_j = \sum_k L'_k (O^{-1})_{kj}$ gives a new set of localizers, and it follows that the equation

$$\mathrm{d}D = \{\mathrm{D}, H\}\mathrm{d}t + \sum_k (L'_k - \langle L'_k \rangle) D \mathrm{d}w'_k \tag{9.26}$$

is the same equation as (9.24) in a different representation. To preserve the density diffusion equations, an orthogonal transformation of the fluctuations must be complemented by an inverse orthogonal transformation of the localizers L_j. Since $\mathrm{d}w'_k$ is just a set of orthonormalized differential fluctuations, they could be replaced by the original $\mathrm{d}w_k$, which shows that the density diffusion equations are independent of the representation of the localizers in L-space. The transformation theory of localizers is similar to the corresponding transformation theory for the master equation in section 4.1, which also applies to QSD.

The QSD theory of localization in position space presented in chapter 5 can be taken over into density diffusion theory. In the laboratory, the density of a particle such as an electron may be partly in a vacuum and partly absorbed in one of a number of solid objects. We can suppose that the vacuum and the objects divide position space into regions k. Frequently the form of the density within each region is of no importance and we are only interested in the probability that the particle will be found in region k, given by integrating the position space density over the whole region. For simplicity we can then assume that the localizers are constant within each region, and can be represented by linear combinations of projectors onto the regions.

Such a localization process takes place in position space, but the phase space theory is just as simple and more general, and so we derive the diffusion equations for a discrete set of quantum weights, each given by an integral of the phase space density over the region k,

$$W_k = \int_\Omega P_k D, \tag{9.27}$$

where P_k is the characteristic function or projector which is 1 inside and 0 outside the region k. The notation highlights the parallel with the corresponding theory for projectors and channels in QSD, which is treated in chapter 5. The expectation of the projector is

$$\langle P_k \rangle = \int_\Omega D P_k = W_k. \tag{9.28}$$

Start with a simple wide open system with a single localizer given by a linear combination of projectors with constant coefficients ℓ_k:

$$L = \sum_k \ell_k P_k, \tag{9.29}$$

whose expectation is

$$\langle L \rangle = \sum_k \ell_k W_k, \tag{9.30}$$

with the density diffusion equation

$$\mathrm{d}D = (L - \langle L \rangle) D \mathrm{d}w = \sum_k \ell_k (P_k - W_k) D \mathrm{d}w. \tag{9.31}$$

The change of the weight $\mathrm{d}W_k$ is then

$$\begin{aligned} \mathrm{d}W_k &= \int_\Omega P_k \mathrm{d}D = \int_\Omega P_k (L - \langle L \rangle) D \mathrm{d}w = \Big(\ell_k - \sum_{k'} \ell_{k'} W_{k'}\Big) W_k \mathrm{d}w \\ &= (\ell_k - \langle \ell \rangle) W_k \mathrm{d}w, \end{aligned} \tag{9.32}$$

where by definition

$$\langle \ell \rangle = \langle L \rangle = \sum_{k'} \ell_{k'} W_{k'}. \tag{9.33}$$

The equations generalize directly to the case of many localizers L_j as

$$dW_k = \sum_j (\ell_{kj} - \langle \ell_j \rangle) W_k dw_j, \tag{9.34}$$

which is the stochastic diffusion equation for the weights.

The special case of one localizer for each region applies, for example, to the case of position space localization. In that case we can label the localizers with k. They are

$$L_k = \ell_k P_k \qquad \text{so} \qquad \ell_{kj} = \delta_{kj}\ell_k \tag{9.35}$$

and the density diffusion equations become

$$dW_k = \left(\ell_k dw_k - \sum_{k'} \langle \ell_{k'} \rangle dw_{k'}\right) W_k. \tag{9.36}$$

Notice that the Hamiltonian evolution has not been included in this generalization. A particular case with Hamiltonian evolution included is given in section 9.9.

9.7 Density entropy decreases

Entropy is a natural measure for the spread of a probability distribution ρ of classical systems in phase space, and is also used to measure the spread of the quantum density D. The theory is simpler than the theory of quantum dispersion entropy of sections 5.6 and 5.7.

The density dispersion entropy S_D of a density D is defined by (9.10) as

$$S_D = -\int_\Omega D \ln D. \tag{9.37}$$

To obtain changes in S_D with time, we need changes in the integrand, which requires the expansion of $\ln D$ up to second order in D. From (5.34) it is given by

$$d(D \ln D) = (1 + \ln D)dD + (dD)^2/(2D). \tag{9.38}$$

Using M $dD = 0$ and the orthonormality of the fluctuations, it follows that

$$\begin{aligned}\mathrm{M}\,\mathrm{d}S_{\mathrm{D}} &= -\int_{\Omega}(\mathrm{d}D)^2/(2D) = -\frac{1}{2}\int_{\Omega}\sum_j (L_j - \langle L_j\rangle)^2 D\mathrm{d}t \\ &= -\frac{1}{2}\sum_j \sigma^2(L_j)\mathrm{d}t = -\frac{1}{2}\sigma^2(L)\mathrm{d}t,\end{aligned} \tag{9.39}$$

where $\sigma^2(L)$ is the variance of the vector localizer $L = \{L_j\}$.

The Hamiltonian evolution does not change the density along the classical paths, making no contribution to the change in the entropy of the density, in contrast to QSD. So unless the variances are all zero, and the density is confined to a subspace of the phase space in which all the L_j are constant, the entropy of the density always *decreases*. This is the density entropy theorem.

Since localization tends to concentrate the density in the phase space, it is not surprising that the entropy always decreases as a result of the localization. This is contrary to the usual idea as to how an entropy should behave. It is one of the remarkable properties of the observed localization of matter in a quantum measurement. Since the probability distribution ρ remains constant along the Hamiltonian trajectories, the fine-grained distribution entropy

$$S = -\int_{\Omega}\rho \ln \rho \tag{9.40}$$

remains constant, so, as discussed in more detail in section 5.9, we cannot use the decrease in entropy to gain useful work, because the loss of entropy of the densities for each member of the ensemble is cancelled by the increase in the distribution entropy which results from the fluctuations. The net result is a constant distribution entropy S. The second law of thermodynamics is about ρ and not about D.

If there is dissipative interaction with the environment, which we do not consider here, the distribution entropy does not remain constant. It normally increases, but can decrease if the environment is at a lower temperature.

For a system with a Hamiltonian, with chaotic regions of phase space and with weak localization, the entropy density theorem can be deceptive. According to the theorem, the regions of phase space in which the density is large decrease in volume, but according to chaos theory, they become stretched, folded and squeezed. As a result, the regions of high density may be small, but they are dispersed over much of the chaotic region. Consequently, a coarse-grained density may remain unlocalized.

It is for this reason that the localization theorem of the next section is for wide open systems, in which the effect of the Hamiltonian is negligible.

9.8 Localization for wide open systems

Suppose there are many L_j, and let G, F be arbitrary dynamical variables, not necessarily connected with the L_j in any way. By using the same methods as in section 9.5, it is easy to show that conditions Co1–Co3 are satisfied for wide open systems when there are many localizers.

For Co4 it is more complicated. For an arbitrary G,

$$\mathrm{d}\langle G\rangle = \int_\Omega G\mathrm{d}D = \sum_j \int_\Omega G(L_j - \langle L_j\rangle)\mathrm{d}w_j = \sum_j \sigma(G, L_j)\mathrm{d}w_j. \tag{9.41}$$

The mean change in the covariance $\sigma(G, F)$ is

$$\begin{aligned}\mathrm{M}\,\mathrm{d}\sigma(\mathrm{G}, \mathrm{F}) &= -\mathrm{M}\,\mathrm{d}\langle G\rangle\mathrm{d}\langle F\rangle = -\mathrm{M}\sum_{kj}\sigma(G, L_j)\mathrm{d}w_j\sigma(G, L_k)\mathrm{d}w_k \\ &= -\mathrm{M}\sum_j \sigma(G, L_j)\sigma(F, L_j)\mathrm{d}t.\end{aligned} \tag{9.42}$$

Notice that the mean is retained in the last expression, since it is not assumed here that the initial ensemble has only one state.

Two special cases of this result are particularly useful. When $G = F$,

$$\mathrm{M}\,\mathrm{d}\sigma^2(G) = -\mathrm{M}\sum_j (\sigma(G, L_j))^2\mathrm{d}t, \tag{9.43}$$

which is minus a sum of squares. Any dynamical variable therefore localizes, unless its covariance with *all* the localizers is zero. Specializing further, suppose L' is a localizer, and $G = F = L'$. In a basis in which $L' = L_1$, we have

$$\mathrm{M}\,\mathrm{d}\sigma^2(L') = -\mathrm{M}\sum_j \sigma(L', L_j)^2\mathrm{d}t \leq -(\mathrm{M}\,\sigma^2(L'))^2\mathrm{d}t. \tag{9.44}$$

The ensemble variance $\Sigma^2(\sigma^2(L'))$ is never negative, so

$$\mathrm{d}(\mathrm{M}\,\sigma^2(L')) = \mathrm{M}\,\mathrm{d}\sigma^2(L') \leq -\mathrm{M}\,(\sigma^2(L'))^2\mathrm{d}t \leq -(\mathrm{M}\,\sigma^2(L'))^2\mathrm{d}t, \tag{9.45}$$

The change in the localization $\Lambda = \Lambda(L')$ of L' is

$$\mathrm{d}\Lambda^{-1}/\mathrm{d}t \leq -\Lambda^2 \qquad \text{and} \qquad \mathrm{d}\Lambda/\mathrm{d}t \geq 1, \tag{9.46}$$

so

$$\Lambda(t) \geq \Lambda(0) + t. \tag{9.47}$$

There is a similar result for any localizer $L' = \sum_j c_j L_j$ with $\sum_j |c_j|^2 = 1$.

When the localization increases, the variance decreases. For a system that is open but not wide open, the variance of a dynamical variable can be increased as a result of the Hamiltonian evolution, unlike the entropy of the density. For example, a density D of a free particle with a spread of momentum and no localizer has a position variance that increases linearly with time. Chaotic Hamiltonians can increase variances exponentially.

So for systems with nonzero Hamiltonians and localizers there is often a competition between them. When the localization is sufficiently strong, the density localizes to a small region around a phase point which then moves in phase space like the phase point of the simple Hamiltonian system. When the localization is very weak, the result depends on the Hamiltonian. Except for special cases, for example when there is a single localizer, which is conserved by the Hamiltonian, integrable systems slowly localize. Chaotic systems were discussed above in section 9.7.

9.9 Localization of a particle in a medium

A particle with an extended wave packet strikes an absorbent solid surface. We know from experiment that in an ensemble of identical wave packets, different wave packets are absorbed at different times. This section uses density diffusion theory to analyse this simple example.

The medium can be a solid, liquid, gas or vacuum. The particle is the system and the medium is the environment, which is treated as uniform, so the rate of localization is independent of the position of the particle within it. For the vacuum it is supposed to be zero. For many different media, each medium occupies a volume labelled k, which includes the whole of the momentum space and a restricted part of position space. The theory of section 9.6 can be used, in which the localization into different regions is expressed in terms of weights W_k. The localization rate can depend on the momentum of the particle, but for simplicity we will neglect this dependence, so the momentum dependence of the distribution plays no role. We can use

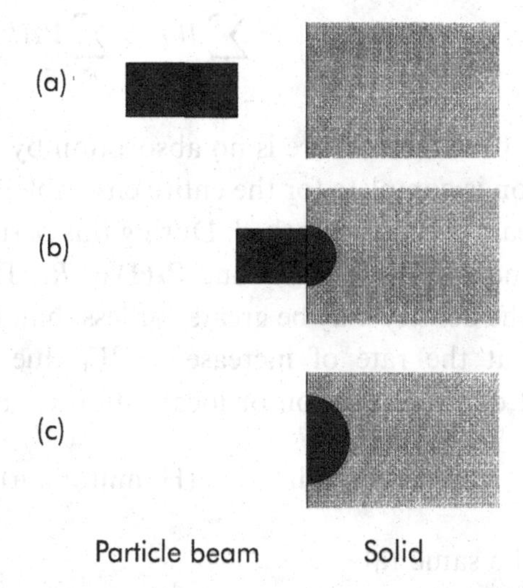

Figure 9.2 A truncated one-particle beam striking a solid surface. The hatched regions sketch the distribution ρ: (a) before, (b) during, and (c) after entry.

densities $D(x, y, z)$, regions k, and the corresponding weights W_k that depend on position alone.

Suppose that a single system of the ensemble is a wave packet in the form of a truncated uniform beam of length L and velocity v, containing a single particle. Suppose that the velocity is in the x direction. The beam strikes a solid surface perpendicular to its motion at time $t = 0$, which absorbs and localizes it in a finite time T. The distribution ρ for the ensemble is sketched in figure 9.2.

We want to find the properties of the quantum weights W_k. Suppose that initially, before it strikes the surface, the density D of each particle of the ensemble occupies the whole beam uniformly, so it is then the same as the distribution ρ. Label the regions of position space by $k = 0$ for the vacuum and $k = 1$ for the solid.

The y and z coordinates play no role, and we consider only the dynamics in the x direction. Denoting space integrals by $\int dx$, the ensemble probabilities $\mathbf{Pr}(k)$ and the individual particle weights W_k are

$$\mathbf{Pr}(0) = \int_{\text{beam}} dx \cdot \rho, \qquad W_0 = \int_{\text{beam}} dx \cdot D,$$
$$\mathbf{Pr}(1) = \int_{\text{solid}} dx \cdot \rho, \qquad W_1 = \int_{\text{solid}} dx \cdot D, \tag{9.48}$$

with the normalization properties

$$\mathrm{M}\, W_k = \mathbf{Pr}(k), \qquad \sum_k W_k = \sum_k \mathbf{Pr}(k) = 1. \tag{9.49}$$

During stage (a) of the figure there is no absorption by the solid, and during stage (c) absorption is complete for the entire ensemble. The interesting stage is (b), when the beam is being absorbed. During this period, $\mathbf{Pr}(1)$ increases at a rate $R = 1/T$, and so $\mathrm{d}\mathbf{Pr}(1) = R\mathrm{d}t$ and $\mathbf{Pr}(1) = Rt$. The weight W_1 for the absorbed part of the density may be greater or less than $\mathbf{Pr}(1)$, but the density is uniform, so that the rate of increase of W_1 due to the Hamiltonian dynamics without density diffusion or localization is given by

$$\mathrm{d}W_1 = W_1 R\mathrm{d}t \qquad \text{(Hamiltonian).} \tag{9.50}$$

W_0 decreases at the same rate.

We will show that the weights are nearly always close to 0 and 1, and change suddenly from one to the other. That is, the weights jump. The dispersion of the weights W_0 and W_1 is conveniently measured by the quantum variances of the projectors P_0, P_1, which are the same, and are given by

$$\sigma^2(P_k) = \langle P_k^2 \rangle - \langle P_k \rangle^2 = W_1 - W_1^2 = W_1 W_0. \tag{9.51}$$

This is small for strong localization, when the weights are close to 0 or 1, larger otherwise. Before $t = 0$ and after $t = T$ the quantum variance $\sigma^2(P_k)$ is zero. We now show that the ensemble mean of this variance is very small at all times, so the weight W_0 jumps suddenly from near 1 to near 0, and conversely for W_1.

Since the density diffusion is uniform in the solid, it can be represented by

$$\begin{aligned} L(x) &= \ell_1 \qquad \text{(solid)}, \\ L(x) &= 0 \qquad \text{(beam)}, \end{aligned} \tag{9.52}$$

where ℓ_1^2 is a rate which is determined by interactions of the particle within the solid, and is therefore very fast by comparison with the rate R.

From equation (9.52) and the density diffusion equation (9.11), the change in the weights due to density diffusion in the solid is therefore

$$\mathrm{d}W_1 = (\ell_1 - \langle \ell \rangle) W_1 \mathrm{d}w = \ell_1 (W_1 - W_1^2)\, \mathrm{d}w \qquad \text{(diffusion).} \tag{9.53}$$

Adding the Hamiltonian and diffusion contributions we get

$$\mathrm{d}W_1 = W_1 R\mathrm{d}t + \ell_1(W_1 - W_1^2)\mathrm{d}w \qquad \text{(total)}, \tag{9.54}$$

which is the equation of change for the weights. Because of the normalization condition in (9.49) there is no need for a separate equation for W_0.

The change in the mean of the quantum variance is

$$\begin{aligned} \mathrm{M}\,\mathrm{d}\sigma^2(\mathrm{P_k}) &= \mathrm{M}\,(\mathrm{d}W_1 - 2W_1\mathrm{d}W_1 - (\mathrm{d}W_1)^2) \\ &= R\mathrm{d}t - 2\mathrm{M}\,W_1 R\mathrm{d}t - \ell_1^2 \mathrm{M}\,(\sigma^2(P_k))^2\mathrm{d}t \\ &\le R\mathrm{d}t - \ell_1^2\mathrm{M}\,(\sigma^2(P_k))^2\mathrm{d}t \\ &\le \left[R - \ell_1^2(\mathrm{M}\,\sigma^2(P_k))^2\right]\mathrm{d}t, \end{aligned} \tag{9.55}$$

where the ensemble variance is used in the last inequality, as usual. The Hamiltonian term tends to increase the mean variance, and the density diffusion term tends to decrease it. Replacing the inequality by an equality gives the boundary, which is obtained analytically, giving

$$\begin{aligned} \mathrm{M}\,\sigma^2 &\le (R/\ell_1^2)^{\frac{1}{2}}\frac{1 - e^{-2k't}}{1 + e^{-2k't}} \qquad (k' = \ell_1 R^{\frac{1}{2}}) \\ &< (R/\ell_1^2)^{\frac{1}{2}} = (\ell_1^2 T)^{-\frac{1}{2}}. \end{aligned} \tag{9.56}$$

Since ℓ_1^2 is a rate determined by the interactions of the particles within the solid, which is much faster than the rate at which the beam enters the solid, $\mathrm{M}\,\sigma^2$ is always much less than unity. So by equation (9.51), W_1 is nearly always very close to 0, its value at $t = 0$, or to 1, its value at $t = T$. Thus it must change very quickly from 0 to 1, a good approximation to an instantaneous jump.

For a single system of the ensemble, the particle remains in the beam and then jumps suddenly into the solid. This is what happens when a quantum particle with an extended matter wave strikes a solid surface, which is justified by a verbal argument in the usual interpretation of quantum mechanics. Here it is derived from the dynamics of density diffusion, as an approximation to quantum state diffusion.

This theory applies generally to systems in which the interaction of the particle with a medium or measuring apparatus is fast, for example in a measurement of a state of a quantum system. Since localization in laboratory quantum measurements is on a macroscopic scale, this shows by a simple example how the quantum jumps which appear in quantum measurements can be derived from quantum state diffusion. It is also an example of localization for which the wave properties of the quantum system are unimportant.

9.10 Discussion

Quantum systems appear to spread out like waves and then become localized like particles. In quantum state diffusion this localization is a stochastic physical process in which the states of individual quantum systems obey a nonlinear QSD equation and the state of an ensemble of quantum systems obeys a linear master equation. The quantum systems of QSD have both wave and localization properties.

Density diffusion theory is an approximation to QSD in which the wave properties are neglected. It is a good approximation when the localization is on a macro scale, and when the quantum system is interacting so strongly with its environment that its coherence no longer has any significant effect and its wave properties are almost completely lost. Such strong interactions occur when signals from quantum systems are amplified and when they are recorded, as in the formation of a photographic latent image [74]. They may also result from interaction of a quantum system with a heat bath, but this is represented by the classical theory of non-Hermitian Lindblads, which has not been treated here.

The state of a quantum system is represented by a density in phase space, with a nontrivial combination of Hamiltonian and localization processes. This is a new kind of nonlinear classical dynamics, which is different for regular and chaotic classical systems, and whose properties are not yet fully worked out. When momentum dependence is relatively unimportant, the process of localization in ordinary position space can be visualized and analysed, as in the example of a particle in a beam striking a solid surface presented in section 9.9.

Density diffusion theory provides a theoretical tool for the study of the macro scale dynamics of localization in quantum systems, and its relative simplicity opens up the possibility of studying more complicated or more subtle localization processes.

10

Semiclassical theory and linear dynamics

The semiclassical limit of QSD is subtle, since it depends on the phase space dispersion of the wave packet. There are two regimes, for if it is large then localization dominates, and if it is small then the localized wave packet follows an approximately classical path in the phase space. The first section describes the relevant classical theory of open systems, and the next two reformulate quantum state diffusion as a semiclassical expansion of the evolution of the Wigner phase space function. The general semiclassical theory of localization in section 10.4 includes a derivation of the localization theory of the previous chapter. Sections 10.5 and 10.6 develop the theory for systems with linear dynamics in terms of localized Gaussian wave packets, or shifted squeezed states. This provides a basis for a more general theory of the wave packet regime, but a completely general theory is not given.

10.1 Classical equations for open systems

The classical theory is needed for comparison. For a system with d freedoms this is a generalization of the theory of Brownian motion of chapter 2, in which the classical trajectories are in phase space. It is convenient to use $2d$-dimensional phase space vectors

$$x = (q, p) = (x_1, \ldots, x_{2d}) = (q_1, \ldots, q_d, p_1, \ldots, p_d) \qquad (10.1)$$

with gradient vectors and partial derivatives

$$\left(\frac{\partial}{\partial \mathrm{q}_1},\ldots,\frac{\partial}{\partial \mathrm{q}_\mathrm{d}},\frac{\partial}{\partial \mathrm{p}_1},\ldots,\frac{\partial}{\partial \mathrm{p}_\mathrm{d}}\right)=\left(\frac{\partial}{\partial q},\frac{\partial}{\partial p}\right)$$
$$=\left(\frac{\partial}{\partial x_1},\ldots,\frac{\partial}{\partial x_{2d}}\right)=(\partial_1,\ldots,\partial_{2d}). \tag{10.2}$$

At first reading it is helpful to think of the simplest example with $d = 1$, so that the phase space has only two dimensions, and the indices r, s that label phase space coordinates have only two values, representing a single coordinate and its conjugate momentum.

Following equation (2.20) for Brownian motion in position space, the general Itô equation in phase space is

$$\mathrm{d}x_r = v_r \mathrm{d}t + \sum_s b_{rs}\mathrm{d}w_s \qquad (r, s = 1, \ldots, 2d), \tag{10.3}$$

where v_r are the drift terms, b_{rs} is the phase space diffusion matrix and $\mathrm{d}w_s$ are the real orthonormal fluctuations introduced in section 2.9. The resultant classical Fokker-Planck equation for the phase space probability distribution $\rho(x_r, t)$ is

$$\dot{\rho} = -\sum_r \partial_r(v_r\rho) + \tfrac{1}{2}\sum_{rs}\partial_r\partial_s((bb^{\mathrm{tr}})_{rs}\rho), \tag{10.4}$$

where b^{tr} is the transposed matrix [125, 55].

10.2 Semiclassical theory of ensembles

The semiclassical theory of this chapter is largely based on [149]. As usual, the limit as $\hbar \to 0$ depends on what is held constant, and this depends on the physics. Here it is supposed that the effect of the environment is held constant, whilst a semiclassical limit is taken for the system. This is not the same as taking the semiclassical limit for system and environment, and only then dividing the system from the environment.

Another subtlety concerns measurement. Quantum measurement, unlike classical measurement, almost always affects the system. Physically, the classical limit of a master equation for quantum measurement can produce no effect on the system, and we will show that this follows from the theory. However, when wave packets are spread out over macro distances or momenta, things are very different for the classical limit of the QSD equation. The solutions of the equation must show localization of phase space densities, both for measurement, as in the previous chapter, and dissipative processes. This chapter shows that there are two different semiclassical limits, depending on the dispersion of the wave in phase space. The limit for the localization regime, when the dispersion is much larger than Planck cells of volume $(2\pi\hbar)^d$, is different from the limit for the wave packet regime, when the dispersion is comparable to a Planck cell. It does not treat the subtle intermediate regime for which the dispersion for different pairs of conjugate variables is in different regimes.

The general semiclassical theory follows from the theory of elementary wide open systems with zero Hamiltonian and one Lindblad given here. The classical limit depends on the relation between commutators and Poisson brackets, which is clarified by rewriting the master equation (4.2) as

$$\begin{aligned}\dot{\rho} &= \mathbf{L}\rho\mathbf{L}^\dagger - \tfrac{1}{2}\mathbf{L}^\dagger\mathbf{L}\rho - \tfrac{1}{2}\rho\mathbf{L}^\dagger\mathbf{L} \\ &= \tfrac{1}{2}\Big([\mathbf{L}\rho, \mathbf{L}^\dagger] - [\rho\mathbf{L}^\dagger, \mathbf{L}]\Big) && \text{(general } \mathbf{L}) \\ &= \tfrac{1}{2}[[\mathbf{L}, \rho], \mathbf{L}] && \text{(Hermitian } \mathbf{L} \text{ only).}\end{aligned} \tag{10.5}$$

In order to obtain a finite classical limit, there must be an explicit $\hbar$-dependence of the Lindblad in equation (10.5). Given that $(-i/\hbar)[\mathbf{F}, \mathbf{G}] \approx \{F, G\}$, the new equations with this dependence are

$$\begin{aligned}\dot{\rho} &= \frac{1}{2\hbar}\Big([\mathbf{L}\rho, \mathbf{L}^\dagger] - [\rho\mathbf{L}^\dagger, \mathbf{L}]\Big) && (\mathbf{L} \to \mathbf{L}/\hbar^{\frac{1}{2}}) && \text{(general } \mathbf{L}), \\ \dot{\rho} &= \frac{1}{2\hbar}[[\mathbf{L}, \rho], \mathbf{L}] && (\mathbf{L} \to \mathbf{L}/\hbar^{\frac{1}{2}}) && \text{(Hermitian } \mathbf{L}).\end{aligned} \tag{10.6}$$

For general Lindblads the change in the density operator ρ for the ensemble has a nonzero classical limit, but for Hermitian Lindblads it does not, so a semiclassical expansion is needed to obtain the first nonzero term. An Hermitian Lindblad has a negligible effect on the change of the density operator in the classical limit. Later we will see that for the individual states of the ensemble this is not true, and the Hermitian Lindblads may even be more important than the Hamiltonian.

The link between quantum and classical is achieved by expressing the quantum master and QSD equations in terms of Wigner functions in phase space. The Wigner-Moyal transform of the density operator ρ is the quantum phase space function

$$\rho_{\hbar}(q,p) = (2/\pi\hbar)^{d/2}\int \mathrm{d}^d q' \, \langle q - q'|\rho|q+q'\rangle\langle q'|2p\rangle, \tag{10.7}$$

which is real because ρ is Hermitian. By convention this is written as the Wigner function

$$W = W(q,p) = \rho_{\hbar}(q,p). \tag{10.8}$$

From this point we will be expanding $\dot{W}$ up to first order in $\hbar$ and expressing it in terms of the zeroth order in the expansion of the Wigner-Moyal Lindblad $L_{\hbar}$, which is just the phase space function $L = L(q,p)$, and so the suffix $\hbar$ can be dropped. The expansion depends on whether or not the operator $\mathbf{L}$ is Hermitian.

The master equation for the density operator can be transformed into a master equation for the Wigner function. The expansion of the resulting change in W up to first order in $\hbar$ is

$$\begin{aligned}
\dot{W} &= \tfrac{1}{2}i(\{LW, L^*\} - \{L^*W, L\}) && \text{(drift)}\\
&\quad -\frac{\hbar}{4}(\{L,\{W,L^*\}\} + \{L^*,\{W,L\}\}) && \text{(general } \mathbf{L}\text{, first order in } \hbar\text{)},\\
\dot{W} &= \tfrac{1}{2}\hbar\{L,\{W,L\}\} && \text{(Hermitian } \mathbf{L}\text{, first order in } \hbar\text{)},
\end{aligned} \tag{10.9}$$

where $\{F, G\}$ is the Poisson bracket. For Hermitian $\mathbf{L}$, the zeroth-order drift is identically zero.

When $\mathbf{L}$ is not Hermitian, $L(q,p)$ can be separated into its real and imaginary parts as

$$L(q,p) = L_{\mathrm{R}}(q,p) + iL_{\mathrm{I}}(q,p). \tag{10.10}$$

Comparison of (10.9) for the zeroth-order term in the change of the Wigner phase space function W with (10.4) for the change in the phase space distribution ρ gives the classical drift term

$$v = (L_{\mathrm{I}}\partial L_{\mathrm{R}}/\partial p - L_{\mathrm{R}}\partial L_{\mathrm{I}}/\partial p,\ L_{\mathrm{R}}\partial L_{\mathrm{I}}/\partial q - L_{\mathrm{I}}\partial L_{\mathrm{R}}/\partial q), \tag{10.11}$$

where the first part represents the drift in the coordinates, and the second part the drift in the conjugate momenta. The change in the distribution for a phase point moving with this drift velocity is given by the divergence of the velocity, proportional to a Poisson bracket:

$$\operatorname{div} v = \nabla \cdot v = -2\{L_{\mathrm{R}}, L_{\mathrm{I}}\}. \tag{10.12}$$

If this is zero, as for Hermitian **L** and some other cases too, Liouville's theorem is satisfied and the phase space flow is incompressible. Otherwise, for a typical dissipative process, the bracket is positive, and the phase space distribution tends to localize in phase space.

An example for $d = 1$ is dissipation due to the annihilation operator of section 5.4, for which

$$L = \gamma^{\frac{1}{2}}(q + ip) \qquad \text{and} \qquad \{L_{\mathrm{R}}, L_{\mathrm{I}}\} = \gamma. \tag{10.13}$$

Obviously this type of localization of distributions is very different from the phase space localization of densities due to quantum measurement, the subject of the last chapter. In the example, it is classical cooling due to a zero-temperature environment.

According to the classical theory, the density goes on localizing indefinitely, so that eventually it occupies a region smaller than a Planck cell, which is unphysical. The classical theory breaks down before this stage is reached. The localization due to the drift is cancelled by the diffusion, which is given by the next term in the semiclassical expansion (10.9). This is an example where terms which appear at first sight to be of different order in Planck's constant have the same order of magnitude, because the density approaches the wave packet regime, and can no longer be considered as independent of Planck's constant.

The $2d \times 2d$ semiclassical diffusion matrix b_{rs} is expressed in terms of two vectors, each with d identical elements:

$$L_{\mathrm{R}}^{d} = (L_{\mathrm{R}}, L_{\mathrm{R}}, L_{\mathrm{R}}, \ldots, L_{\mathrm{R}}, L_{\mathrm{R}}), \qquad L_{\mathrm{I}}^{d} = (L_{\mathrm{I}}, L_{\mathrm{I}}, L_{\mathrm{I}}, \ldots, L_{\mathrm{I}}, L_{\mathrm{I}}). \tag{10.14}$$

It is

$$b_{rs} = \sqrt{\hbar/d}\,(-\partial/\partial p, \partial/\partial q)^{\mathrm{tr}}\,(L_{\mathrm{R}}^{d}, L_{\mathrm{I}}^{d}), \tag{10.15}$$

which gives a quantum diffusion proportional to $\hbar$. Notice that the Hermitian and anti-Hermitian parts of the Lindblad have a classical interpretation. The example represents zero-temperature dissipation, which must be accompanied by fluctuations, because of the quantum fluctuation-dissipation theorem. In this case the fluctuations only become significant when the phase space dispersion approaches the size of a Planck cell.

For Hermitian $\mathbf{L}$, the zeroth-order drift is zero, and the phase space density is preserved for the drift, so Liouville's theorem is satisfied. The dissipation is of order $\hbar$, and of course all imaginary parts $\mathbf{L}_\mathrm{I}$ are zero.

As usual, the general theory can be obtained by adding in a Hamiltonian and summing over many Lindblads. The result is

$$\left.\begin{aligned} \dot{W} &= \{H, W\} + \tfrac{1}{2} i \sum_i (\{L_i W, L_i^*\} - \{L_i^* W, L_i\}) \\ &\quad - \frac{\hbar}{4} \sum_i (\{L_i, \{W, L_i^*\}\} + \{L_i^*, \{W, L_i\}\}) \quad \text{(general)}, \\ \dot{W} &= \{H, W\} - \tfrac{1}{2}\hbar \sum_i \{L_i, \{W, L_i\}\} \qquad \text{(all } \mathbf{L}_i \text{ Hermitian)}, \end{aligned}\right\} \tag{10.16}$$

for the semiclassical limits of the master equations 10.6.

10.3 Semiclassical theory of pure states

This follows from the semiclassical limit of the QSD equation. The projector form (4.30) of the QSD equation is closest to the ensemble analysis of the previous section, so it is used here. The pure state equivalent of the ensemble master equation (10.1) is the equation for the change in the projector $\mathbf{P} = \mathbf{P}_\psi$, which is

$$\begin{aligned} & \mathrm{d}\mathbf{P} = (\mathbf{L}\mathbf{P}\mathbf{L}^\dagger - \tfrac{1}{2}\mathbf{L}^\dagger\mathbf{L}\mathbf{P} - \tfrac{1}{2}\mathbf{P}\mathbf{L}^\dagger\mathbf{L})\mathrm{d}t + 2\mathrm{Re}\mathbf{L}_\Delta\mathbf{P}_\psi\mathrm{d}\xi, \\ \text{so} \quad & \mathrm{d}\mathbf{P} = \tfrac{1}{2}\Big([\mathbf{L}\mathbf{P}, \mathbf{L}^\dagger] - [\mathbf{P}\mathbf{L}^\dagger, \mathbf{L}]\Big)\mathrm{d}t + 2\mathrm{Re}\mathbf{L}_\Delta\mathbf{P}_\psi\mathrm{d}\xi \qquad \text{(general } \mathbf{L}\text{)}, \\ \text{and} \quad & \mathrm{d}\mathbf{P} = \tfrac{1}{2}[[\mathbf{L}, \mathbf{P}], \mathbf{L}]\mathrm{d}t + 2\mathrm{Re}\mathbf{L}_\Delta\mathbf{P}_\psi\mathrm{d}\xi \qquad \text{(Hermitian } \mathbf{L}\text{)}, \end{aligned} \tag{10.17}$$

where 2Re represents the sum of the operator and its Hermitian conjugate and operates on all factors to its right.

Since the drift of the projector QSD equation (4.30) is the same as for the Fokker-Planck equation, we only have to look in detail at the contribution of the fluctuation terms. Let d_fl represent a change due to the fluctuations alone, omitting the drift terms which are proportional to $\mathrm{d}t$.

For consistency, the same factors of powers of $\hbar$ must be used for the fluctuations as for the drift, and, with the definitions of $\mathbf{L}$ in (10.6),

$$\mathrm{d}_\mathrm{fl}\mathbf{P} = \hbar^{-\frac{1}{2}} 2\mathrm{Re}\mathbf{L}_\Delta\mathbf{P}_\psi\mathrm{d}\xi \qquad \text{(all } \mathbf{L}\text{)}. \tag{10.18}$$

Using $P = P_\hbar(q, p)$ for the Wigner phase space representation of the pure state projector $\mathbf{P}_\psi$, the contribution of the fluctuation to the variation of P is

$$\mathrm{d}_{\mathrm{fl}}P = 2\mathrm{Re}\,(\hbar^{-1/2}L_\Delta P\mathrm{d}\xi + (i/2)\,\hbar^{1/2}\{L, P\}\mathrm{d}\xi) \tag{10.19}$$

to order $\hbar^{\frac{1}{2}}$, where

$$L_\Delta(q, p) = L(q, p) - \int_\Omega L(q, p)P(q, p). \tag{10.20}$$

From (10.16) and (10.19), the equations for $\mathrm{d}P$ with a Hamiltonian and any number of Lindblads are

$$\begin{aligned}\mathrm{d}P = {} & \{H, P\}\mathrm{d}t + \tfrac{1}{2}i\sum_i(\{L_iP, L_i^*\} - \{L_i^*P, L_i\})\mathrm{d}t \\ & - \frac{\hbar}{4}\sum_i(\{L_i, \{P, L_i^*\}\} + \{L_i^*, \{P, L_i\}\})\mathrm{d}t \\ & + 2\mathrm{Re}\sum_i(\hbar^{-1/2}L_{i\Delta}P\mathrm{d}\xi_i + (i/2)\,\hbar^{1/2}\{L_i, P\}\mathrm{d}\xi_i) \quad \text{(general)},\end{aligned} \tag{10.21a}$$

$$\begin{aligned}\mathrm{d}P = {} & \{H, P\}\mathrm{d}t - \tfrac{1}{2}\hbar\sum_i\{L_i, \{P, L_i\}\}\mathrm{d}t \\ & + 2\mathrm{Re}\sum_i(\hbar^{-1/2}L_{i\Delta}P\mathrm{d}\xi_i + (i/2)\,\hbar^{1/2}\{L_i, P\}\mathrm{d}\xi_i) \quad (\text{all } \mathbf{L}_i \text{ Hermitian}).\end{aligned} \tag{10.21b}$$

The striking thing about these equations is that the classical limit is singular, with a singularity of order $\hbar^{-1/2}$, even for Hermitian $\mathbf{L}$, when the classical effect on the density operator is zero. The physical implications of this are discussed in the next section.

10.4 Localization regime

In the localization regime the dispersion is much larger than a Planck cell.

For an elementary system with a single Hermitian $\mathbf{L}$, (10.21b) becomes

$$\mathrm{d}P = \{H, P\}\mathrm{d}t + 2\,\hbar^{-1/2}L_\Delta P\mathrm{d}\xi_{\mathrm{R}} - \hbar^{1/2}\{L, P\}\mathrm{d}\xi_{\mathrm{I}} \qquad (\text{order } \hbar^{\frac{1}{2}}), \tag{10.22}$$

where $d\xi_R$, $d\xi_I$ are the real and imaginary parts of $d\xi$, which are both normalized to $\frac{1}{2}dt$. When $\mathbf{L}_\Delta$ is of order $\hbar^0$, so that the variance $\mathbf{L}$ is of classical size, the second term dominates. It is the density diffusion equation (9.16) with $D = P$ and a different definition of L. The system is effectively wide open because the effect of the Hamiltonian is of lower order than the effect of the Lindblad. For this regime the density diffusion theory of the previous chapter can be taken over in its entirety, for a single Lindblad or many Lindblads.

Dynamical variables proportional to $L(q,p)$ localize. As a result

$$\mathrm{M}\,d\sigma^2(L) = \mathrm{M}\,d\langle(L_\Delta)^2\rangle \leq 0 \tag{10.23}$$

and the magnitude of L_Δ tends to decrease. As L localizes, the singular term becomes relatively less important, and the other terms become more significant. At first the Hamiltonian starts to have an effect, so the projector tends to follow the Hamiltonian flow in phase space. The term in $\hbar^{\frac{1}{2}}$ is also like a Hamiltonian flow, but it is proportional to the fluctuation $d\xi_I$ and not to dt. Consequently the motion is dominated by this fluctuation term for sufficiently small time intervals, and by the usual Hamiltonian term for longer time intervals.

The localization regime of a system with a general $\mathbf{L}$ is similar. Both the real and imaginary parts of the Lindblad tend to localize, as we saw for the simple example of the annihilation operator in section 5.4. However, for general $\mathbf{L}$ the term

$$\{LP, L^*\} - \{L^*P, L\} \tag{10.24}$$

is significant. When $[\mathbf{L}_R, \mathbf{L}_I] \neq 0$, then $\{L_R, L_I\} \neq 0$ and this term prevents the wave packet from localizing into a region of phase space for which Heisenberg indeterminacy would be violated.

The localization proceeds until the size of the wave packet is affected by these additional terms, which is the wave packet regime of the next two sections.

10.5 Linear phase space transformations and squeezed states

This section and the next apply QSD to systems whose dynamics is linear, in particular systems whose Hamiltonian is quadratic and whose Lindblads are linear in coordinates and momenta. It has no special connection with the linear QSD of section 4.5. It has close connections with the moving basis of chapter 6, and is needed for the theory of the wave packet regime. A related study of squeezed states is in [150], of semiclassical mechanics is in [92] and of Bogolubov transformations is in [14].

We start with a completely localized elementary system of one freedom for which the initial state of the system $|\psi(0)\rangle$ at time $t=0$ is a minimum indeterminacy wave packet, which is a Gaussian. For these states and others like them, Planck's constant $\hbar$ is the natural unit of action, and it is put equal to unity here, so that a Planck cell has area 2π. For the initial state, the coordinate frame (q, p) in phase space can be shifted and scaled by a canonical transformation to a frame (q_0, p_0) so that $|\psi(0)\rangle$ is the null eigenstate of the annihilation operator $\mathbf{a}_0 = (1/\sqrt{2})(\mathbf{q}_0 + i\mathbf{p}_0)$, giving

$$\mathbf{a}_0|\psi(0)\rangle = 0, \qquad |\psi(0)\rangle = |0. + i0.\rangle \qquad \text{(initial state)}, \qquad (10.25)$$

where the eigenvalue $0. + i0.$ of the annihilation operator is used to label the state. This state is also the ground state $|0\rangle$ of the Hamiltonian $\mathbf{H} = \frac{1}{2}(\mathbf{q}^2 + \mathbf{p}^2)$, one of the Fock number states $|n\rangle$ labelled by an integer. Of course, the expectation value of $\mathbf{a}_0$ for this state is

$$\langle\psi(0)|\mathbf{a}_0|\psi(0)\rangle = \langle\mathbf{a}_0\rangle = 0. + i0., \qquad (10.26)$$

so the real and imaginary parts, which are the expectation values of the position and momentum operators, are zero too.

This section treats shifting and scaling in detail.

The definition of the annihilation operator depends on the choice of canonical coordinates for the phase space. For new coordinates (q', p'), there will be a new annihilation operator $\mathbf{a}'$ and a new null eigenstate of that annihilation operator. The null eigenstate for a given (q', p') will be called the standard wave packet for that coordinate system.

For complex $\alpha = \alpha_R + i\alpha_I$, a *coherent state* $|\alpha\rangle$, centred at the phase point $(q, p) = \sqrt{2}(\alpha_R, \alpha_I)$ is produced by shifting the wave packet from the origin, producing a new eigenstate of the annihilation operator:

$$\mathbf{a}_0|\alpha\rangle = \alpha|\alpha\rangle \qquad \text{(coherent state)}. \qquad (10.27)$$

Obviously $|\alpha\rangle$ is the null eigenstate of a shifted annihilation operator

$$\mathbf{a}_\alpha = \mathbf{a}_0 - \alpha\mathbf{I}, \qquad (10.28)$$

which can be considered as the annihilation operator associated with a shifted origin in the phase space. Its null eigenstate is the standard wave packet for the coordinates with the shifted origin.

A shift in phase space is a special example of a general linear canonical transformation. To each of these general transformations there corresponds a transformed annihilation operator whose null eigenstate is a general *squeezed state* or *Gaussian wave packet* in the original representation.

In quantum theory, linear changes of coordinate frame are special changes of representation, and here the classical frame changes will be called by the same name. Linear transformations of coordinates and momenta look just the same whether they are transformations of quantum operators $(\mathbf{q}, \mathbf{p})$ or classical phase space variables (q, p). Both of them are simpler in terms of the quantum annihilation operator

$$\mathbf{a} = (1/\sqrt{2})(\mathbf{q} + i\mathbf{p}) \qquad \text{(annihilation, } \hbar = 1) \tag{10.29}$$

or its classical equivalent, which is the classical complex amplitude or *amplitude*

$$a = (1/\sqrt{2})(q + ip) \qquad \text{(amplitude, } \hbar = 1). \tag{10.30}$$

Notice that the quantum unit of action is used for both.

We need to represent canonical transformations in terms of amplitudes a. For one freedom, though not for more, a canonical transformation from (q, p) to $(q'p')$ is defined by the condition that the area is preserved by the transformation, so that the Jacobian of the transformation is unity:

$$(\partial q/\partial q')(\partial p/\partial p') - (\partial q/\partial p')(\partial p/\partial q') = 1 \qquad \text{(area preservation).} \tag{10.31}$$

This condition applied to the complex variables a and ia^* shows that they are canonical conjugates, which is to be expected, since the corresponding operators obey the standard quantum commutation relations for canonically conjugate operators. A general linear canonical transformation can be defined in terms of a general amplitude a that is a linear function of the initial amplitude a_0 of (10.25) and its conjugate a_0^*:

$$a = -\alpha + \beta a_0 + \gamma^* a_0^*, \tag{10.32}$$

where α, β and γ are all complex numbers. The α here is a shift, just as it was above, and the sign is chosen for consistency with (10.28). The inverse of this transformation has coefficients α_-, β_-, γ_- given by

$$\alpha_- = \alpha^*\gamma^* - \beta\alpha, \qquad \beta_- = \beta, \qquad \gamma_- = -\gamma. \tag{10.33}$$

The complex conjugate of (10.32) shows that

$$ia^* = -i\alpha^* + i\gamma a_0 + i\beta^* a_0^* \tag{10.34}$$

and the condition that a and ia^* must satisfy the area-preserving condition (10.31) gives

$$|\beta|^2 = 1 + |\gamma|^2 \qquad \text{(canonical)}, \tag{10.35}$$

so, for a canonical transformation, $|\beta| \geq 1$. The null eigenstate of the operator

$$\mathbf{a} = -\alpha\mathbf{I} + \beta\mathbf{a}_0 + \gamma^*\mathbf{a}_0^* \tag{10.36}$$

corresponding to (10.32) defines a Gaussian wave packet or squeezed state, giving a correspondence

transformation $\rightarrow$ amplitude a $\rightarrow$ operator $\mathbf{a}$ $\rightarrow$ wave packet

between canonical transformations and Gaussian wave packets. There is a unique standard wave packet for every α, β, γ defining a canonical transformation. In the old representation the wave packet is a shifted squeezed state, but in the new representation it is the standard wave packet.

There is not a *unique* transformation for a given Gaussian wave packet. A null eigenstate of an annihilation operator $\mathbf{a}$ is also an eigenstate of $u\mathbf{a}$ where u is a complex phase factor of unit modulus, corresponding to a rotation about the origin of the complex plane or the phase space, and to the substitutions

$$\alpha \rightarrow u\alpha, \qquad \beta \rightarrow u\beta, \qquad \gamma \rightarrow u^*\gamma. \tag{10.37}$$

So we can always choose a phase so that β is positive. This condition makes the transformation unique, and also defines β uniquely in terms of γ, using (10.35):

$$\beta = \sqrt{1 + |\gamma|^2}. \tag{10.38}$$

An arbitrary Gaussian wave packet or squeezed state is defined uniquely as the standard wave packet of a coordinate frame defined by (α, γ). It is the null eigenstate $|\alpha, \gamma\rangle$ of the transformed annihilation operator $\mathbf{a}(\alpha, \gamma)$:

$$\mathbf{a}(\alpha, \gamma)|\alpha, \gamma\rangle = 0,$$

$$\text{so} \qquad (-\alpha\mathbf{I} + \beta\mathbf{a}_0 + \gamma^*\mathbf{a}_0^\dagger)|\alpha, \gamma\rangle = 0. \tag{10.39}$$

Note that $\mathbf{a}(\alpha, 0. + i0.)$ annihilates a coherent state shifted by $\sqrt{2}\,\alpha_R$ in the q-direction and $\sqrt{2}\,\alpha_I$ in the p-direction.

It is helpful to have a geometric picture of the wave packets. If the standard wave packet $|0. + i0., 0. + i0.\rangle$ is represented by a circle of area 2π at the origin, then $|\alpha, \gamma\rangle$ is represented by an ellipse of the same area, centred at $\sqrt{2}\,(\alpha_R, \alpha_I)$, with the ratio of its axes determined by the magnitude of γ and the direction of its major axis given by the phase of γ.

We will show in the next section that the QSD equation for an elementary linear system turns Gaussian wave packets into Gaussian wave packets. They are preserved. Consequently the solution can be expressed as an evolving state $|\alpha(t), \gamma(t)\rangle$ which can be pictured as an ellipse of area 2π, whose centre follows a classical trajectory in the phase space determined by $\alpha(t)$ and whose shape is determined by $\gamma(t)$. Typically the ellipse changes its eccentricity and orientation with time.

This classical picture of completely localized states complements the classical picture of localization presented in the previous chapter. The next section obtains the stochastic equations for $|\alpha(t), \gamma(t)\rangle$.

10.6 Linear dynamics and the linear approximation

Linear systems are simpler to analyse than others, and were used as examples by the early researchers on QSD. This section does not contain a full treatment of QSD for linear systems, and can be supplemented by looking at the papers of Halliwell and Zoupas [81, 161] and the numerical example in [21].

Consider first a simple wide open system with an initial Gaussian wave packet at $t = 0$ and a Lindblad linear in $\mathbf{q}$ and $\mathbf{p}$. Choose the initial representation (q_0, p_0) so that the initial wave packet is a standard wave packet:

$$|\psi(0)\rangle = |0\rangle = |0. + i0., 0. + i0.\rangle, \tag{10.40}$$

where an integer labels a Fock state. Let $\mathbf{a}_0$ be the annihilation operator for this representation, which is used as the standard representation to which others are related. Since the initial Lindblad $\mathbf{L}_0$ is linear, it can be written as

$$\mathbf{L}_0 = b\mathbf{a}_0 + c^*\mathbf{a}_0^\dagger, \tag{10.41}$$

where b is real and c is complex, and terms proportional to the unit operator are ignored because they have the same effect as a Hamiltonian. Consequently $\langle 0|\mathbf{L}_0|0\rangle = 0$ and so from (4.20) the state at time $\mathrm{d}t$ is

$$\begin{aligned}|\psi'\rangle &= |\psi(0)\rangle - \tfrac{1}{2}\mathbf{L}_0^\dagger\mathbf{L}_0|\psi(0)\rangle\mathrm{d}t + \mathbf{L}_0|\psi(0)\rangle\mathrm{d}\xi \\ &= |0\rangle - \tfrac{1}{2}(c\mathbf{a}_0 + b\mathbf{a}_0^\dagger)c^*\mathbf{a}_0^\dagger|0\rangle\mathrm{d}t + c^*\mathbf{a}_0^\dagger|0\rangle\mathrm{d}\xi \\ &= |0\rangle - \tfrac{1}{2}|c|^2|0\rangle\mathrm{d}t - (1/\sqrt{2})bc^*|2\rangle\mathrm{d}t + c^*|1\rangle\mathrm{d}\xi.\end{aligned} \tag{10.42}$$

For the solution $|\psi'\rangle$ take a trial Gaussian wave packet

$$|\psi'\rangle = |\alpha', \gamma'\rangle = |\mathrm{d}\alpha, \mathrm{d}\gamma\rangle, \tag{10.43}$$

which is the null eigenstate of the annihilation operator $\mathbf{a}(\mathrm{d}\alpha, \mathrm{d}\gamma)$, so that from (10.39), in the original representation

$$\begin{aligned}0 &= \mathbf{a}(\mathrm{d}\alpha, \mathrm{d}\gamma)|\psi'\rangle \\ &= [-\mathrm{d}\alpha\mathbf{I} + (1 + \mathrm{d}\beta)\mathbf{a}_0 + \mathrm{d}\gamma^*\mathbf{a}_0^\dagger] \\ &\quad \times [|0\rangle - \tfrac{1}{2}|c|^2|0\rangle\mathrm{d}t - bc^*(1/\sqrt{2})|2\rangle\mathrm{d}t + c^*|1\rangle\mathrm{d}\xi] \\ &= (-\mathrm{d}\alpha + (1 + \mathrm{d}\beta)c^*\mathrm{d}\xi)|0\rangle \\ &\quad + (-c^*\mathrm{d}\alpha\mathrm{d}\xi - bc^*\mathrm{d}t + \mathrm{d}\gamma^*)|1\rangle \\ &\quad + (c^*\sqrt{2}\,\mathrm{d}\gamma^*\mathrm{d}\xi)|2\rangle,\end{aligned} \tag{10.44}$$

where negligible products of differentials have been ignored. This complicated equation has the simple solution

$$\mathrm{d}\alpha = c^*\mathrm{d}\xi, \qquad \mathrm{d}\gamma = bc\mathrm{d}t. \tag{10.45}$$

So the state at time $\mathrm{d}t$ is the Gaussian wave packet

$$|\psi(\mathrm{d}t)\rangle = |c^*\mathrm{d}\xi, bc\mathrm{d}t\rangle \tag{10.46}$$

and linear dynamics preserves Gaussian wave packets. The centroid of the wave-packet has a Brownian motion in the phase space. Over short times the shape of the wave packet drifts. Over longer periods it is affected by the fluctuations. So for an ensemble of identical Gaussian wave packets, the individual states diffuse apart in the phase space, like Brownian particles moving in two dimensions, but at first they change their shape together.

Now suppose the equations have been integrated up to time t, with a known state vector $|\alpha(t), \gamma(t)\rangle$, which is annihilated by $\mathbf{a}(\alpha(t), \gamma(t))$, and known values $\alpha_-(t)$, $\gamma_-(t)$ for the inverse transformation.

The Lindblad can be time-dependent, with original representation

$$\mathbf{L}_0(t) = b(t)\mathbf{a}_0 + c^*(t)\mathbf{a}_0^\dagger. \tag{10.47}$$

The transformation defined by $\alpha(t)$, $\gamma(t)$ can be used to transform $\mathbf{L}_0(t)$ to the current representation. The equations can then be integrated for time $\mathrm{d}t$ as above, getting the new state in the current representation, which can then be transformed back to the original representation using $\alpha_-(t)$, $\gamma_-(t)$.

For a Hamiltonian and two Lindblads, the changes $\mathrm{d}\alpha$, $\mathrm{d}\gamma$ over a time $\mathrm{d}t$, produced by the Hamiltonian and by each Lindblad separately, can be added together, by the additivity rule. Any more linear Lindblads are linearly dependent, so they are equivalent to two. The generalization to many freedoms is more complicated, and is not treated here.

The *linear approximation* is the application of the linear theory to nonlinear systems. It requires the time-dependent operators corresponding to the phase functions obtained by expanding

$$\partial H(q,p)/\partial q, \qquad \partial H(q,p)/\partial p \quad \text{and} \quad L_j(q,p) \tag{10.48}$$

about the current centroid of the wave packet given by

$$(q(t), p(t)) = \sqrt{2}(\alpha_\mathrm{R}(t), \alpha_\mathrm{I}(t)) \tag{10.49}$$

and then using the linear theory to integrate forward by a time δt. The approximation ignores any deviations from Gaussian wave packets, and can be used when localization is nearly complete and the higher powers of the expansion of (10.48) are relatively small over the region occupied by the wave packets.

The linear approximation provides a good picture of the localization regime for the semiclassical theory, with approximate Gaussian wave packets whose position and shape in phase space change with time. The general theory of the classical limit in the localization regime and its connection with the usual theory of open classical systems is subtle. It is treated using coordinates and momenta in [149].

10.7 Summary and discussion

The two regimes of the semiclassical theory have very different properties. For a sufficiently small ratio of Planck's constant to the relevant classical variables with the dimensions of action, localization dominates in the localization regime. As the relative value of $\hbar$ increases, the effects of the Hamiltonian begin to appear, and then the fluctuations of Hamiltonian form. As expected, the classical theory appears in the limit $\hbar \to 0$, but the details of this limit are not simple. Care is needed when drawing physical conclusions from these limits for intrinsic fluctuations like primary state diffusion, because these depend on the magnitude of the Lindblads, which can be very small compared with the other operators.

When the semiclassical theory is applicable, then localization continues until the Hamiltonian and Lindblad phase functions look linear on the scale of the wave packet. The linear theory is then a good approximation, the states are represented by Gaussians in either position or momentum representation, and the evolution of a single system can be expressed in terms of an evolving stochastic change in a linear canonical representation. The wave packet can be pictured in phase space as an ellipsoid with the volume of a Planck cell whose position fluctuates and whose shape also changes, like squeezed states of arbitrary orientation in quantum optics. The motion of the centroid of the wave packet can be considered as a classical property, but the shape of the wave packet is a quantum property. Nevertheless, the theory and computation of the evolution of the wave packet depend only on the classical mechanics of the linear system. As an approximation to a nonlinear system, the theory depends only on the behaviour of phase space functions in the neighbourhood of the trajectory of the centroid of the wave packet.

References

[1] D.Z. Albert. On quantum mechanical automata. *Phys. Lett. A*, 98:249–252, 1983.

[2] L. Allen and J. Eberly. *Optical Resonance and Two-level Atoms*. Dover, New York, 1st edition, 1987.

[3] A. Aspect, J. Dalibard and G. Roger. Experimental test of Bell inequalities using time-varying analysers. *Phys. Rev. Lett.*, 49:1804–1807, 1982.

[4] A. Barchielli and V.P. Belavkin. Measurements continuous in time and a-posteriori states in quantum mechanics. *J. Phys. A*, 24:1495–1514, 1991.

[5] A. Barenco, T.A. Brun, R. Schack and T.P. Spiller. Effects of noise on quantum error correction algorithms. *Phys. Rev. A*, 56:1177–1188, 1997.

[6] H.-P. Bauer and F. Petruccione. Stochastic dynamics of reduced wave functions and continuous measurement in quantum optics. *Fortschr. Phys.*, 45:39–78, 1997.

[7] D. Bedford and D. Wang. Towards an objective interpretation of quantum mechanics. *Nuo. Cim.*, B26:313–315, 1975.

[8] D. Bedford and D. Wang. A criterion for state vector reduction. *Nuo. Cim.*, B37:55–62, 1977.

[9] V.P. Belavkin, O. Hirota and R.L. Hudson, eds, *Quantum Communications and Measurement*, Plenum, New York, 1995.

[10] J.S. Bell. *Speakable and Unspeakable in Quantum Mechanics*. Cambridge University Press, 1987.

[11] P. Benioff. The computer as a physical system. *J. Stat. Phys.*, 22:563–591, 1980.

[12] P. Benioff. Quantum mechanical models of Turing machines that dissipate no energy. *Phys. Rev. Lett.*, 48:1581, 1982.

[13] P. Benioff. Quantum mechanical Hamiltonian models of computers. In D.M. Greenberger, ed., *New Techniques and Ideas in Quantum Measurement Theory*, New York Academy of Sciences, New York, 1986.

[14] J-P. Blaizot and G. Ripka. *Quantum Theory of Finite Systems*. MIT Press, Cambridge, Mass. 1986.

[15] D. Bohm. *Quantum Theory*. Prentice-Hall, New York, 1951.

[16] D. Bohm. *Causality and Chance in Modern Physics*. Routledge and Kegan Paul, London, 1984.

[17] D. Bohm and J. Bub. *Rev. Mod. Phys.*, 38:453–469, 1966.

[18] D. Bohm and B.J. Hiley. *The Undivided Universe*. Routledge, London, 1993.

[19] V.B. Braginsky and F.Ya. Khalili. *Quantum Measurement*. Cambridge University Press, 1992.

[20] T.A. Brun and N. Gisin. Quantum state diffusion and time correlation functions. *J. Mod. Opt.*, 43:2289–2300, 1996.

[21] T.A. Brun, I.C. Percival and R. Schack. Quantum chaos in open systems: a quantum state diffusion analysis. *J. Phys. A*, 29:2077–2090, 1996.

[22] H.J. Carmichael. *An Open Systems Approach to Quantum Optics*. Springer, Berlin, 1993.

[23] C.M. Caves. Quantum limits on noise in linear amplifiers. *Phys. Rev. D*, 26:1817, 1982.

[24] C. Cohen-Tannoudji and J. Dalibard. Single-atom laser spectroscopy – looking for dark periods in flourescent light. *Europhys. Lett.*, 1:441, 1986.

[25] C. Cohen-Tannoudji, B. Zambon and E. Arimondo. Quantum-jump approach to dissipative processes: application to amplification without inversion. *J. Opt. Soc. Am. B*, 10:2107–2120, 1993.

[26] E.B. Davies. *Quantum Theory of Open Systems*. Academic, London, 1976.

[27] H. Dehmelt. Continuous Stern–Gerlach effect – principle and idealized apparatus. *Proc. Nat. Acad. Sci. USA*, 83:2291, 1986.

[28] H. Dehmelt. Continuous Stern-Gerlach effect – noise and the measurement process. *Proc. Nat. Acad. Sci. USA*, 83:3074, 1986.

[29] H.G. Dehmelt. *Bull. Am. Phys. Soc.*, 20:60, 1975.

[30] H.G. Dehmelt. mono-ion oscilator as potential laser frequency standard. *IEEE Trans. Instrum. Meas.*, 31:83, 1982.

[31] D. Deutsch. Quantum theory, the Church–Turing principle and the universal quantum computer. *Proc. R. Soc. Lond. A*, 400:97–117, 1985.

[32] L. Diósi. Stochastic pure state representation for open quantum systems. *Phys. Lett. A*, 114:451–454, 1986.

[33] L. Diósi. A universal master equation for the gravitational violation of quantum mechanics. *Phys. Lett.*, 120:377–381, 1987.

[34] L. Diósi. Continuous quantum measurement and Itô formalism. *Phys. Lett. A*, 129:419–423, 1988.

[35] L. Diósi. Localized solution of a simple nonlinear quantum Langevin equation. *Phys. Lett. A*, 132:233–236, 1988.

[36] L. . Quantum stochastic processes as models for state vector reduction. *J. Phys. A*, 21:2885–2898, 1988.

[37] L. Diósi Models for universal reduction of macroscopic quantum fluctuations. *Phys. Rev. A*, 40:1165–1174, 1989.

[38] L. Diósi and B. Lukács. Calculation of X-ray signals from Károlyházy hazy spacetime. *Nuo. Cim.*, 108B:1419–1421, 1993.

[39] L. Diósi and B. Lukács. Károlyházy's quantum spacetime generates neutron star density in vacuum. *Nuo. Cim.*, 108B:1419–1421, 1993.

[40] L. Diósi, N. Gisin, J. Halliwell and I.C. Percival. Decoherent histories and quantum state diffusion. *Phys. Rev. Lett.*, 74:203–207, 1995.

[41] P.A.M. Dirac. Evolution of the physicist's picture of nature. *Sci. American*, 208:45, 1963.

[42] H.F. Dowker and J.J. Halliwell. The quantum mechanics of history: the decoherence functional in quantum mechanics. *Phys. Rev. D*, 46:1580–1609, 1992.

[43] P.D. Drummond, K.J. McNeil and D.F. Walls. Nonequilibrium transitions in second harmonic generation i. semiclassical theory. *Optica Acta*, 27:321, 1980.

[44] P.D. Drummond, K.J. McNeil and D.F. Walls. Nonequilibrium transitions in second harmonic generation ii. quantum theory. *Optica Acta*, 28:211, 1981.

[45] R. Dum, P. Zoller and H. Ritsch. Monte Carlo simulation of the atomic master equation for spontaneous emission. *Phys. Rev. A*, 45:4879–4887, 1992.

[46] A. Einstein. *Investigations on the Theory of the Brownian Movement*. Dover, New York, 1956.

[47] A. Einstein, B. Podolsky and N. Rosen Can quantum-mechanical description of reality be considered complete? *Phys. Rev.*, 47:777–780, 1935.

[48] J. Ellis, S. Mohanty and D.V. Nanopoulos. Quantum gravity and the collapse of the wave function. *Phys. Lett.*, 221B:113–119, 1989.

[49] H. Everett. Relative state formulation of quantum mechanics. *Rev. Mod. Phys.*, 29:454, 1957.

[50] R.P. Feynman, F.B. Moringo and W.G. Wagner. *Lectures on Gravitation*. California Institute of Technology, 1973.

[51] R.P. Feynman. Simulating physics with computers. *Int. J. Theor. Phys.*, 21:467, 1982.

[52] R.P. Feynman. Quantum mechanical computers. *Found. Phys.*, 16:507–531, 1986.

[53] S. J. Freeman J.F. Clauser. *Phys. Rev. Lett.*, 28:938, 1972.

[54] C.W. Gardiner, A.S. Parkins and P. Zoller. Wave-function quantum stochastic differential equations and quantum-jump simulation methods. *Phys. Rev. A*, 46:4363–4381, 1992.

[55] C.W. Gardiner. *Handbook of Stochastic Methods*. Springer, Berlin, 2nd edition, 1985.

[56] C.W. Gardiner. *Quantum Noise*. Springer, Berlin, 1st edition, 1991.

[57] B.M. Garraway and P.L. Knight. Evolution of quantum superpositions in open environments: Quantum trajectories, jumps and localization in phase space. *Phys. Rev. A*, 50:2548–2563, 1994.

[58] B.M. Garraway, P.L. Knight, and J. Steinbach. Dissipation of quantum superpositions – localization and jumps. *Appl. Phys. B*, 60:63–68, 1995.

[59] B.M. Garraway and P.L. Knight. Comparison of quantum-state diffusion and quantum-jump simulations of two-photon processes in a dissipative environment. *Phys. Rev. A*, 49:1266–1274, 1994.

[60] M. Gell-Mann and J.B. Hartle. Quantum mechanics in the light of quantum cosmology. In W.H. Zurek, ed., *Complexity, Entropy and the Physics of Information*, Addison Wesley, Redwood City, CA, 1990.

[61] M. Gell-Mann and J.B. Hartle. Classical equations for quantum systems. *Phys. Rev. D*, 47:3345–3382, 1993.

[62] G.-C. Ghirardi, A. Rimini and T. Weber. Disentanglement of quantum wave functions: Answer to comment on 'unified dynamics for microscopic and macroscopic systems'. *Phys. Rev. D*, 36:3287–3289, 1986.

[63] G.-C. Ghirardi, A. Rimini and T. Weber. Unified dynamics for microscopic and macroscopic systems. *Phys. Rev. D*, 34:470–491, 1986.

[64] G.-C. Ghirardi, P. Pearle and A. Rimini. Markov processes in Hilbert space and continuous spontaneous localization of systems of identical particles. *Phys. Rev. A*, 42:78–89, 1990.

[65] G.-C. Ghirardi, R. Grassi and P. Pearle. Relativistic dynamical reduction models: General framework and examples. *Foundations of Physics*, 20:1271–1316, 1990.

[66] G.-C. Ghirardi, R. Grassi and A. Rimini. A continuous spontaneous reduction model involving gravity. *Phys. Rev. A*, 42:1057–1064, 1990.

[67] N. Gisin. Comment on 'quantum measurements and stochastic processes'. *Phys. Rev. Lett.*, 53:1775–1776, 1984.

[68] N. Gisin. Quantum measurements and stochastic processes. *Phys. Rev. Lett.*, 52:1657, 1984.

[69] N. Gisin. Stochastic quantum dynamics and relativity. *Helvetica Physica Acta*, 62:363–371, 1989.

[70] N. Gisin. Weinberg's nonlinear quantum mechanics and supraluminal communications. *Phys. Lett. A*, 143:1–2, 1990.

[71] N. Gisin. Time correlations and Heisenberg picture in the quantum state diffusion model of open systems. *J. Mod. Opt.*, 40:2313–2319, 1993.

[72] N. Gisin and I.C. Percival. The quantum state diffusion model applied to open systems. *J. Phys. A*, 25:5677–5691, 1992.

[73] N. Gisin and I.C. Percival. Quantum state diffusion, localisation and quantum dispersion entropy. *J. Phys. A*, 26:2233–2243, 1993.

[74] N. Gisin and I.C. Percival. The quantum state diffusion picture of physical processes. *J. Phys. A*, 26:2245–2260, 1993.

[75] N. Gisin and I.C. Percival. Quantum state diffusion, from foundations to applications. In R.S. Cohen, ed., *Experimental Metaphysics*. Kluwer, Dordrecht, 1997.

[76] N. Gisin and M. Rigo. Relevant and irrelevant Schrödinger equations. *J. Phys. A*, 28:7375–7390, 1995.

[77] N. Gisin, T.A. Brun and M. Rigo. From quantum to classical: the quantum state diffusion model. In M. Ferrero and A. van der Merwe, eds, *New Developments on Fundamental Problems in Quantum Physics*, pages 141–149, Kluwer, Dordrecht, 1997.

[78] P. Goetsch and R. Graham. Quantum trajectories for nonlinear optical processes. *Ann. Physik*, 2:706–719, 1993.

[79] P. Goetsch and R. Graham. Linear stochastic wave equations for continuously measured quantum systems. *Phys. Rev. A*, 50:5242–5255, 1994.

[80] R. Griffiths. Consistent histories and the interpretation of quantum mechanics. *J. Statist. Phys.*, 36:219–272, 1984.

[81] J.J. Halliwell and A. Zoupas. Quantum state diffusion, density matrix diagonalization and decoherent histories: A model. *Phys. Rev. D*, 52:7294, 1995.

[82] S. Hawking. Unpredictability of quantum gravity. *Comm. Math. Phys.*, 87:395–415, 1982.

[83] M. Holland, S. Marksteiner, P. Marte, and P. Zoller. Measurement-induced localization from spontaneous decay. *Phys. Rev. Lett.*, 76:3683–3686, 1996.

[84] P.R. Holland. *The Quantum Theory of Motion*. Cambridge University Press, 1993.

[85] R.L. Hudson and K.R. Parthasarathy. Quantum Itô's formula and stochastic evolution. *Comm. Math. Phys.*, 93:301–323, 1984.

[86] E. Joos and H.D. Zeh. The emergence of classical properties through interaction with the environment. *Z. Phys. B*, 59:223–243, 1985.

[87] A. Károlyházy, F. Frenkel and B. Lukács. On the possible role of gravity in the reduction of the wave function. In R. Penrose and C.J. Isham, eds, *Quantum Concepts in Space and Time*, pp 109–128. Clarendon, Oxford, 1986.

[88] F. Károlyházy. Gravitation and quantum mechanics of macroscopic objects. *Nuo. Cim.*, A 42:390–402, 1996.

[89] M. Kasevich and S. Chu. Atomic interferometry using stimulated Raman transitions. *Phys. Rev. Lett.*, 67:181–184, 1991.

[90] M. Lax. Quantum noise x: Density-matrix treatment of field and population-difference fluctuations. *Phys. Rev.*, 157:213, 1967.

[91] G. Lindblad. On the generators of quantum dynamical semigroups. *Comm. Math. Phys.*, 48:119–130, 1976.

[92] R.G. Littlejohn. The semiclassical evolution of wave packets. *Phys. Reports*, 138:193–291, 1986.

[93] T. Maudlin. *Quantum Non-locality and Relativity*. Blackwell, Oxford, 1994.

[94] M.B. Mensky. *Continuous Quantum Measurements and Path Integrals*. Institute of Physics, Bristol, 1993.

[95] G.J. Milburn. Intrinsic decoherence in quantum mechanics. *Phys. Rev. A*, 44:5401–5406, 1991.

[96] K. Mølmer, Y. Castin and J. Dalibard. Monte-Carlo wave function method in quantum optics. *J. Opt. Soc. Am. B*, 10:524–538, 1993.

[97] H. Moya-Cessa, V. Buzek, M.S. Kim and P.L. Knight. Intrinsic decoherence in the atom-field interaction. *Phys. Rev. A*, 48:3900–3905, 1993.

[98] E. Nelson. Derivation of the Schrödinger equation from Newtonian mechanics. *Phys. Rev.*, 150:1079–1085, 1966.

[99] E. Nelson. *Dynamical Theories of Brownian Motion*. Princeton University Press, Princeton, 1967.

[100] R. Omnès. Consistent interpretations of quantum mechanics. *Rev. Mod. Phys.*, 64:339–382, 1992.

[101] R. Omnès. *The Interpretation of Quantum Mechanics*. Princeton University Press, Princeton, 1994.

[102] K.R. Parthasarathy. *An Introduction to Quantum Stochastic Calculus*. Birkhäuser, Basel, 1992.

[103] P. Pearle. Reduction of the state vector by a nonlinear Schrödinger equation. *Phys. Rev. D*, 13:857–868, 1976.

[104] P. Pearle. Towards explaining why events occur. *Int. J. Theor. Phys.*, 18:489–518, 1979.

[105] P. Pearle. Comment on ‘quantum measurements and stochastic processes’. *Phys. Rev. Lett.*, 53:1775, 1984.

[106] P. Pearle. Stochastic dynamical reduction theories and superluminal communication. *Phys. Rev. D*, 33:2240–2252, 1986.

[107] P. Pearle. Combining stochastic dynamical state vector reduction with spontaneous localization. *Phys. Rev. A*, 39:2277–2289, 1989.

[108] R. Penrose and C.J. Isham, eds. *Quantum Concepts in Space and Time*. Clarendon, Oxford, 1986.

[109] I.C. Percival. Atomic scattering computations. In P.G. Burke and B.L. Moiseiwitsch, eds, *Atomic Processes and Applications*, pages 321–329, North-Holland, Amsterdam, Holland, 1976..

[110] I.C. Percival. Quantum records. In P. Cvitanović, I. Percival and A. Wirzba, eds, *Quantum Chaos, Quantum Measurement*, pages 199–204, Kluwer, Dordrecht, Holland, 1992.

[111] I.C. Percival. Localization of wide-open quantum systems. *J. Phys. A*, 27:1003–1020, 1994.

[112] I.C. Percival. Primary state diffusion. *Proc. R. Soc. Lond. A*, 447:189–209, 1994.

[113] I.C. Percival. Environmental and primary state diffusion. In V.P. Belavkin, ed., *Quantum Communication and Measurement*, pages 265–280. Plenum, New York, 1995.

[114] I.C. Percival. Quantum spacetime fluctuations and primary state diffusion. *Proc. R. Soc. Lond. A*, 451:503–513, 1995.

[115] I.C. Percival and W.T. Strunz. Atom interferometry for quantum gravity? In M. Ferrero and A. van der Merwe, eds, *New Developments on Fundamental Problems in Quantum Physics*, pages 291–300, Kluwer, Dordrecht, 1997.

[116] I.C. Percival. Atom interferometry, spacetime and reality. *Physics World*, 10(3):43–48, March 1997.

[117] I.C. Percival and W.T. Strunz. Detection of spacetime fluctuations by a model matter interferometer. *Proc. R. Soc. Lond. A*, 453:431–446, 1997.

[118] I.C. Percival and W.T. Strunz. Classical dyanamics of quantum localization. *J. Phys. A*, 31:1815–1830, 1998.

[119] A. Peres. *Quantum Theory: Concepts and Methods.* Kluwer, Dordrecht, 1995.

[120] B. Pippard. Physics in 1900. In A. Brown, L.M. Pais and B. Pippard, eds, *Twentieth Century Physics*, volume 1, pages 1–41. Institute of Physics, Bristol, 1995.

[121] M.B. Plenio and P.L. Knight. The quantum jump approach to dissipative dynamics in quantum optics. *Rev. Mod. Phys.*, 70:101–144, 1998.

[122] J. Polchinski. Weinberg's nonlinear quantum mechanics and the Einstein–Podolsky–Rosen paradox. *Phys. Rev. Lett.*, 66:397, 1991.

[123] A.I.M Rae. *Quantum Mechanics.* Institute of Physics, Bristol, 1992.

[124] M. Rigo, F. Mota-Furtado, G. Alber and P.F. O'Mahony. Quantum state diffusion model and the damped driven nonlinear oscillator. *Phys. Rev. A*, 55:1165–1673, 1997.

[125] H. Risken. *The Fokker-Planck Equation.* Springer, Berlin, 1989.

[126] J.L. Sánchez-Gómez. Decoherence through stochastic fluctuations in the gravitational field. In L. Diósi and B. Lucács, eds, *Stochastic evolution of quantum states in open sytems and in measurement processes*, pages 88–93, World Scientific, Singapore, 1994.

[127] R. Schack, T.A. Brun and I.C. Percival. Quantum state diffusion, localization and computation. *J. Phys. A*, 28:5401–5413, 1995.

[128] R. Schack, T.A. Brun and I.C. Percival. Quantum state diffusion with a moving basis: computing quantum-optical spectra. *Phys. Rev. A*, 53:2694, 1996.

[129] R. Schack, G.M. D'Ariano and C.M. Caves. Hypersensitivity to perturbation in the quantum kicked top. *Phys. Rev. E*, 50:972–987, 1994.

[130] R. Schack and T.A. Brun. A C++ library using quantum trajectories to solve quantum master equations. *Comp. Phys. Comm.*, 102:210–228, 1997.

[131] E. Schrödinger. The present situation in quantum mechanics. *Naturwissenschaften*, 23:807–812, 823–828, 844–849, 1935.

[132] E. Schrödinger. The present situation in quantum mechanics. *Proc. Camb. Phil. Soc.* , 31:555, 1935.

[133] E. Schrödinger. The present situation in quantum mechanics. *Proc. Camb. Phil. Soc.*, 32:446, 1936.

[134] A. Shimony. Controllable and uncontrollable nonlocality. In S. Kamefuchi, ed., *Foundations of quantum mechanics in the Light of New Technology*, pages 225–230, Physical Society of Japan, Tokyo, 1984.

[135] P.W. Shor. In S. Goldwasser, ed., *Proceedings of the 35th Annual Symposium on the Theory of Computer Science*, page 124. IEEE Computer Society Press, Los Alamitos, California, 1994.

[136] P.W. Shor. Scheme for reducing decoherence in quantum computer memory. *Phys. Rev. A*, 52:R2493–R2496, 1995.

[137] T.P. Spiller, J.F. Prance, T.D. Clark, R.J. Prance and H. Prance. Behaviour of individual ultra-small capacitance of quantum devices. *Int. J. Mod. Phys. B*, 11:779–803, 1997.

[138] T.P. Spiller. Quantum information processing: cryptography, computation and teleportation. *Proc. IEEE*, 84:1719–1746, 1996.

[139] T.P. Spiller, B.M. Garraway and I.C. Percival. Thermal-equilibrium in the quantum state diffusion picture. *Phys. Lett. A*, 179:63–66, 1993.

[140] T.P. Spiller and J.F. Ralph. The emergence of chaos in an open quantum system. *Phys. Lett. A*, 194:235–240, 1994.

[141] M.D. Srinivas. Quantum theory of continuous measurements and its applications to quantum optics. *Pramana*, 47:1–23, 1996.

[142] M.D. Srinivas and E.B. Davies. Photon counting probabilities in quantum optics. *Optica Acta*, 28:981, 1981.

[143] A.M. Steane. Error-correcting codes in quantum theory. *Phys. Rev. Lett.* , 77:793, 1996.

[144] A.M. Steane. Multiple-particle interference and quantum error correction. *Proc. R. Soc. Lond. A*, 452:2551, 1996.

[145] T. Steimle and G. Alber. The continuous Stern-Gerlach effect and the quantum state diffusion model of state reduction. *Phys. Rev. A*, 53:1282–1291, 1996.

[146] T. Steimle, G. Alber and I.C. Percival. Mixed classical-quantal representation for open quantum systems. *J. Phys. A*, 28:L491–L496, 1995.

[147] J. Steinbach, B.M. Garraway and P.L. Knight. High-order unraveling of master-equations for dissipative evolution. *Phys. Rev. A*, 51:3302–3308, 1995.

[148] S. Stenholm and M. Wilkens. Jumps in quantum theory. *Contemporary Physics*, 38:257–268, 1997.

[149] W.T. Strunz and I.C. Percival. The semiclassical limit of quantum state diffusion – a phase space approach. *J. Phys. A*, 31:1801–1813, 1998.

[150] D.F. Walls and G.J. Milburn. *Quantum Optics*. Springer, Berlin, 1994.

[151] S. Weinberg. Testing quantum mechanics. *Ann. Phys. (N.Y.)*, 194:336, 1989.

[152] S. Weinberg. Precision tests of quantum mechanics. *Phys. Rev. Lett.*, 62:485, 1989.

[153] J.A. Wheeler. Everett's 'relative state' formulation of quantum theory. *Rev. Mod. Phys.*, 29:463, 1957.

[154] J.A. Wheeler and W.H. Zurek, eds. *Quantum Theory and Measurement*. Princeton University Press, Princeton, 1983.

[155] H.M. Wiseman and G.J. Milburn. Interpretation of quantum jump and diffusion processes illustrated on the Bloch sphere. *Phys. Rev. A*, 47:1652–1666, 1993.

[156] H.M. Wiseman and G.J. Milburn. Quantum theory of field quadrature measurements. *Phys. Rev. A*, 47:642–662, 1993.

[157] H.D. Zeh. On the interpretation of measurement in quantum theory. *Found. Phys.*, 1:69, 1970.

[158] H.D. Zeh. There are no quantum jumps, nor are there particles. *Phys. Lett. A*, 172:189–192, 1993.

[159] X.P. Zheng and C.M. Savage. Quantum trajectories and classical attractors in 2nd-harmonic generation. *Phys. Rev. A*, 51:792–797, 1995.

[160] M. Zoller, P. Marte and D.F. Walls. Quantum jumps in atomic systems. *Phys. Rev. A*, 35:198, 1987.

[161] A. Zoupas. Phase-space localization and approach to thermal equilibrium for class of open systems. *Phys. Lett. A*, 219:162, 1996.

INDEX

Page numbers in italic indicate where definitions appear